Prepared in cooperation with the United States Marine Corps

A Water-Budget Model and Estimates of Groundwater Recharge for Guam

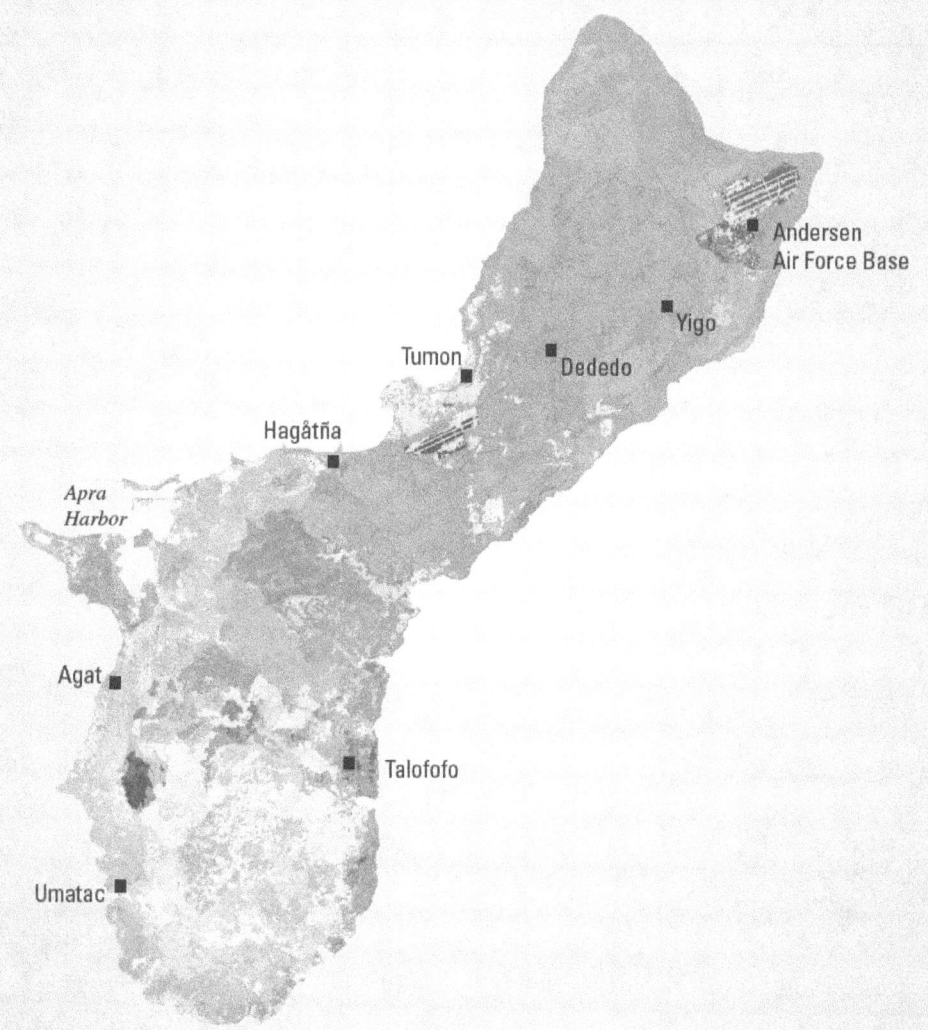

Scientific Investigations Report 2012–5028

U.S. Department of the Interior
U.S. Geological Survey

COVER: Map of Guam showing the relative distribution of estimated annual recharge for baseline conditions. Estimated recharge is shown using a special color scale with red representing very low recharge and dark blue representing very high recharge.

A Water-Budget Model and Estimates of Groundwater Recharge for Guam

By Adam G. Johnson

Prepared in cooperation with the United States Marine Corps

Scientific Investigations Report 2012–5028

U.S. Department of the Interior
U.S. Geological Survey

U.S. Department of the Interior
KEN SALAZAR, Secretary

U.S. Geological Survey
Marcia K. McNutt, Director

U.S. Geological Survey, Reston, Virginia: 2012

This report and any updates to it are available online at:
http://pubs.usgs.gov/sir/2012/5028/

For more information on the USGS—the Federal source for science about the Earth, its natural and living resources, natural hazards, and the environment—visit *http://www.usgs.gov* or call 1–888–ASK–USGS

For an overview of USGS information products, including maps, imagery, and publications, visit *http://www.usgs.gov/pubprod*

To order this and other USGS information products, visit *http://store.usgs.gov*

Any use of trade, product, or firm names is for descriptive purposes only and does not imply endorsement by the U.S. Government.

Suggested citation:
Johnson, A.G., 2012, A water-budget model and estimates of groundwater recharge for Guam: U.S. Geological Survey Scientific Investigations Report 2012–5028, 53 p.

iii

Contents

Figures

Tables

Conversion Factors, Datums, and Abbreviations and Acronyms

Inch/Pound to SI

Multiply	By	To obtain
Length		
inch (in.)	2.54	centimeter (cm)
inch (in.)	25.4	millimeter (mm)
foot (ft)	0.3048	meter (m)
mile (mi)	1.609	kilometer (km)
Area		
acre	4,047	square meter (m^2)
acre	0.4047	hectare (ha)
square foot (ft^2)	0.09290	square meter (m^2)
square mile (mi^2)	259.0	hectare (ha)
square mile (mi^2)	2.590	square kilometer (km^2)
Volume		
gallon (gal)	3.785	liter (L)
gallon (gal)	0.003785	cubic meter (m^3)
million gallons (Mgal)	3,785	cubic meter (m^3)
Flow rate		
million gallons per day (Mgal/d)	0.04381	cubic meter per second (m^3/s)
inch per hour (in/h)	0.0254	meter per hour (m/h)
inch per day (in/d)	25.4	millimeter per day (mm/d)
inch per year (in/y)	25.4	millimeter per year (mm/yr)
mile per hour (mi/h)	0.447	meter per second (m/s)

Temperature in degrees Celsius (°C) may be converted to degrees Fahrenheit (°F) as follows: °F=(1.8×°C)+32

Temperature in degrees Fahrenheit (°F) may be converted to degrees Celsius (°C) as follows: °C=(°F-32)/1.8

Concentrations of chemical constituents in water are given in milligrams per liter (mg/L).

Vertical coordinate information is referenced to mean sea level.

Horizontal coordinate information is referenced to the World Geodetic System of 1984 (WGS 84).

Elevation, as used in this report, refers to distance above the vertical datum.

Abbreviations and Acronyms

AFB	Air Force Base
CDM	Camp, Dresser & McKee Inc.
ET	Evapotranspiration
FAO	Food and Agricultural Organization
FEIS	Final Environmental Impact Statement
GIS	Geographic Information System
GWA	Guam Waterworks Authority
NCDC	National Climatic Data Center
NWQL	National Water Quality Laboratory
UFW	Unaccounted for water
USGS	U.S. Geological Survey
USMC	United States Marine Corps
WERI	Water and Environmental Research Institute
WGS	World Geodetic System

A Water-Budget Model and Estimates of Groundwater Recharge for Guam

By Adam G. Johnson

Abstract

On Guam, demand for groundwater tripled from the early 1970s to 2010. The demand for groundwater is anticipated to further increase in the near future because of population growth and a proposed military relocation to Guam. Uncertainty regarding the availability of groundwater resources to support the increased demand has prompted an investigation of groundwater recharge on Guam using the most current data and accepted methods. For this investigation, a daily water-budget model was developed and used to estimate mean recharge for various land-cover and rainfall conditions. Recharge was also estimated for part of the island using the chloride mass-balance method.

Using the daily water-budget model, estimated mean annual recharge on Guam is 394.1 million gallons per day, which is 39 percent of mean annual rainfall (999.0 million gallons per day). Although minor in comparison to rainfall on the island, water inflows from water-main leakage, septic-system leachate, and stormwater runoff may be several times greater than rainfall at areas that receive these inflows. Recharge is highest in areas that are underlain by limestone, where recharge is typically between 40 and 60 percent of total water inflow. Recharge is relatively high in areas that receive stormwater runoff from storm-drain systems, but is relatively low in urbanized areas where stormwater runoff is routed to the ocean or to other areas. In most of the volcanic uplands in southern Guam where runoff is substantial, recharge is less than 30 percent of total water inflow.

The water-budget model in this study differs from all previous water-budget investigations on Guam by directly accounting for canopy evaporation in forested areas, quantifying the evapotranspiration rate of each land-cover type, and accounting for evaporation from impervious areas. For the northern groundwater subbasins defined in Camp, Dresser & McKee Inc. (1982), mean annual baseline recharge computed in this study is 159.1 million gallons per day, which is 50 percent of mean annual rainfall, and is 42 percent greater than the recharge estimate of Camp, Dresser & McKee Inc. (1982). For the northern aquifer sectors defined in Mink (1991), which encompass most of the northern half of the island, mean annual baseline recharge computed in this study is 238.0 million gallons per day, which is 51 percent of mean annual rainfall, and is about 6 percent lower than the recharge estimate of Mink (1991). For the drought simulation performed in this study, recharge for the entire island is 259.3 million gallons per day, which is 34 percent lower than recharge computed for baseline conditions. For all aquifer sectors defined by Mink (1991), total recharge during drought conditions is 32 percent lower than mean baseline recharge. For the future land-cover water-budget simulation, which represents potential land-cover changes owing to the military relocation and population growth, estimated recharge for the entire island is nearly equal to the baseline recharge estimate that was based on 2004 land cover.

Using the water-budget model, estimated recharge in the northern half of the island is most sensitive to crop coefficients and net precipitation rates—two of the water-budget parameters used in the estimation of total evapotranspiration. Estimated recharge in the southern half of the island is most sensitive to crop coefficients, net precipitation rate, and runoff-to-rainfall ratios.

During March 2010 to May 2011, bulk-deposition samples from five rainfall stations on Guam were collected and analyzed for chloride. Additionally, samples from five groundwater sites were collected and analyzed for chloride. Results were used to estimate groundwater recharge using the chloride mass-balance method. Recharge estimates using this method at three bulk-deposition stations on the northern limestone plateau range from about 25 to 47 percent of rainfall. These recharge estimates are similar to the estimate of Ayers (1981) who also used this method. Recharge estimates at each bulk-deposition station, however, are lower than the baseline recharge estimate from the water-budget model used in this study. This may be because no large storms, such as tropical cyclones, passed near Guam during March 2010 to May 2011.

Introduction

Owing to population growth, freshwater demand on Guam has increased in the past and will likely increase in the future. From 1970 to 2010, the resident population on Guam grew from about 86,000 to 181,000 (U.S. Census Bureau, 2011). In addition, the tourist industry brings many visitors, and Guam hosted just over 1 million tourists in 2010 (Guam Visitors Bureau, 2010). From the early1970s to 2010, groundwater withdrawals from the freshwater lens of northern Guam, the main source of freshwater on the island, tripled from about 15 to 45 million gallons per day (Mgal/d) according to Mink

(1976) and the U.S. Navy, Naval Facilities Engineering Command, Pacific (NAVFAC Pacific, 2010d). In 2010, total freshwater demand on Guam was about 58 Mgal/d, with groundwater supplying nearly 80 percent of that demand (NAVFAC Pacific, 2010d). By 2020, population growth on Guam is projected to add 23,000 people to the 2010 population (U.S. Census Bureau, 2011). In addition to this growth, a proposed military buildup in the near future includes the transfer of about 8,600 United States Marines and their 9,000 dependents from Okinawa, Japan, to Guam, new facilities for berthing of an aircraft carrier, and a new missile defense task force. The total population growth due to the military relocation is projected to be about 34,000 within 10 years of initial relocation actions (NAVFAC Pacific, 2010a). As a result of the military relocation and population growth, freshwater demand on Guam is projected to increase substantially (NAVFAC Pacific, 2010d).

To ensure prudent management of the groundwater resources of Guam, and to plan for sustainable development, an improved understanding of the magnitude and spatial distribution of groundwater recharge is needed. Previous recharge estimates for Guam were very generalized or pertained to only a part of the island. Groundwater recharge estimates from this report, quantified for the entire island and based on the most current data available, will assist in the development of groundwater-flow models of Guam's freshwater-lens system.

Purpose and Scope

This report describes the (1) development of a daily water-budget model for computing groundwater recharge for the entire island of Guam, (2) application of the model to estimate long-term mean annual recharge for various land-cover and rainfall conditions, and (3) data collected for the chloride mass-balance method as an independent estimate of groundwater recharge. Recharge estimates from this study are compared to previously published recharge estimates, and the sensitivity of recharge estimates to selected water-budget parameters is evaluated.

Previous Investigations

Previous water-budget investigations estimated recharge for parts of northern Guam (table 1). One of the earliest recharge estimates was published by Mink (1976). In 1982, Camp, Dresser & McKee Inc. (1982) published The Northern Guam Lens Study in which recharge and sustainable-yield values were estimated for aquifer subbasins that span much of the interior parts of northern Guam. Mink (1991) revised the Camp, Dresser & McKee Inc. (1982) subbasin boundaries and estimated recharge for the northern half of the island. Each of these previous investigations estimated recharge as mean monthly rainfall minus mean monthly evapotranspiration. Recharge estimates from these water-budget investigations may be biased because they use monthly time steps. In contrast, water-budget models that use daily time steps provide a more realistic simulation of short-duration rainfall events and daily evapotranspiration (Oki, 2008). More recently, Jocson and others (2002) and Habana and others (2009) used water budgets that operated on daily time steps to estimate recharge for parts of northern Guam. The water budget presented in this report uses a daily time step and provides estimates of recharge for the entire island.

Using the chloride mass-balance method as an alternative approach to water budgets, Ayers (1981) estimated recharge for the northern half of the island on the basis of the ratio of chloride in rainfall to chloride in groundwater. The recharge estimate of Ayers (1981) may be low because it was based on rainfall samples that were collected during 24-hour periods. Such short-duration sample periods may not allow for adequate accounting of other sources of atmospheric chloride deposition, most notably sea spray. Additionally, rainfall samples collected by Ayers (1981) did not account for seasonal variations in chloride deposition to the land surface. For this study, recharge is estimated using a chloride mass-balance method that attempts to account for all atmospheric and seasonal variations of chloride deposition to the land surface by integrating conditions over longer sampling periods. Samples were collected in 2- to 4-month intervals during March 2010 to May 2011.

Acknowledgments

This study was conducted in cooperation with the United States Marine Corps (USMC) and in collaboration with the University of Guam's Water and Environmental Research Institute of the Western Pacific (WERI). The author thanks John Engott and Delwyn Oki of the U.S. Geological Survey for assistance with the water-budget model. Thanks to Nathan Habana of the Water and Environmental Research Institute of the Western Pacific (WERI); Travis Hylton of Naval Facilities Engineering Command, Pacific; and Brett Railey, Martin Roush, and Rodney Toves of Guam Waterworks Authority (GWA) for providing data for the water-budget model. Tomoko Bell, Vivianna Bendixson, Jennifer Cruz, Jacob Fathal, John Jenson, Mark Lander, Blaz Miklavic, and Kennedy Tolenoa of WERI, Lisa Fiedler of Joint Guam Program Office, Francis Lizama of GWA, and Chip Guard of the National Weather Service office in Guam each provided valuable assistance with chloride sampling.

Description of Guam

Physical Setting

Guam, the largest and southernmost of the Mariana Islands, lies in the tropical western Pacific Ocean between latitude 13°14'N and 13°40'N and between longitude 144°37'E and 144°58'E (fig. 1). The island is divided into northern and southern geographic provinces by the Adelup fault, which extends across the center of the island from Pago Bay to Adelup Point. The land surface, 211 square miles in area, consists of four major physiographic land forms: limestone plateau,

Table 1. Previous water-budget investigations for Guam.

Reference	Area	Time step
Mink (1976)	Most of northern Guam	Monthly
Camp, Dresser & McKee Inc. (1982)	Northern aquifer subbasins	Monthly
Mink (1991)	Northern aquifer sectors	Monthly
Jocson and others (2002)	Parts of Yigo-Tumon and Finegayan aquifer sectors	Daily
Habana and others (2009)	Parts of Yigo-Tumon and Mangilao aquifer sectors	Daily

volcanic uplands, interior basin, and coastal lowlands (Tracey and others, 1964; Young, 1988) (fig. 1).

The southern half of the island is predominantly rugged volcanic uplands that have been cut by streams. A discontinuous ridge of mountains, with many peaks over 1,000 ft in altitude, extends from Mount Chachao to the southern tip of the island. The highest point on Guam, Mount Lamlam at 1,332 ft, is on the narrow band of limestone that caps the volcanic uplands. The steeply dissected terrain west of the mountain ridgeline is footed by coastal lowlands that span from Asan Point to just north of Facpi Point. Orote Peninsula is a limestone plateau up to 200 ft high that extends west of the southern province. East of the mountain ridgeline, the volcanic land surface is characterized by steep slopes at higher elevations and has gently rolling hills at lower elevations. A partially incised limestone plateau fringes the east coast from Pago Bay to Inarajan Bay. The interior basin is a structural depression that extends inland from Talofofo Bay to the dissected limestone cap; it includes Fena Valley Reservoir, the dissected karst terrane northeast of Fena Valley Reservoir, and rolling hills and valleys comprised of volcanic, limestone, and alluvial units. Flat-lying areas occur within lower reaches of stream valleys and at stream mouths along the southern and eastern coastlines.

The northern half of the island is a broad limestone plateau bordered by steep cliffs and discontinuous coastal lowlands. Gently tilted to the southwest, the plateau slopes from an altitude of more than 600 ft in the north to less than 200 ft near the narrow center part of the island northwest of the Adelup fault. Volcanic rocks protrude through the limestone plateau at Mataguac Hill and Mount Santa Rosa. Except for cliffs, the slope of the land surface is generally less than 10 percent. Most of the plateau lacks stream channels, but has many closed depressions. Near the southern volcanic uplands, however, the plateau is cut by drainage channels that funnel runoff into closed depressions and the Hagåtña Swamp.

Climate

Guam is warm and humid throughout the year. At the Guam International Airport, the average monthly air temperature ranges between 80° and 83°F. Relative humidity typically ranges between 65 and 80 percent during the day, and between 85 and 100 percent at night. The spatial variability of average monthly air temperature and dew point temperature across Guam is minor. According to modeled estimates (Daly and Halbleib, 2006a), in any given month, the spatial variability of average minimum temperature and the spatial variability of average maximum temperature across the island are less than 7°F. In any given month, the spatial variability of average dew point temperature across the island is less than 3°F. Localized temperatures that are not resolved in these modeled estimates, however, may be different from these average values.

Seasonal differences in rainfall and wind define the distinct wet and dry seasons on Guam. Across the northern plateau, about one-third of the annual rainfall occurs during the dry season months, (January through June), and about two-thirds during the wet season months (July through December) (Lander and Guard, 2003). During the dry season, northeasterly trade winds are persistent, and most rainfall occurs as light showers in amounts typically no more than 0.25 in. per day (Lander and others, 2001). Mean monthly rainfall across the island is less than 8 in. during the dry season. During the wet season, winds weaken and typically veer to the southeast, and the atmosphere over the island is more humid and unstable (Guard and others, 1999). As a result, rainfall is predominantly of convective origin, and occurs as moderate to heavy downpours (Lander and others, 2001). Mean monthly rainfall across the island ranges between 5 and 18 in. during the wet season. Rainfall producing weather systems during the wet season, range in spatial scale from isolated thunderheads to weather systems that affect the entire island including: (1) clusters of convective clouds, (2) monsoon squall lines, and (3) convective clouds associated with the periphery or core of tropical cyclones (Lander and others, 2001). On average, about 30 percent of the wet season rainfall is induced by tropical cyclones (Kubota and Wang, 2009). The passage of typhoons near or directly over the island can produce torrential downpours with rainfall rates exceeding 6 inches in an hour and 20 inches in 24 hours (Lander and Guard, 2003).

Mean annual rainfall ranges from about 84 in. near Apra Harbor, to about 116 in. in the southern highlands (Daly and Halbleib, 2006b) (fig. 2). The largest deviations from mean rainfall conditions are related to tropical cyclones and El Niño/ La Niña-Southern Oscillation events (Lander and Guard, 2003). Some of the wettest years have occurred during years

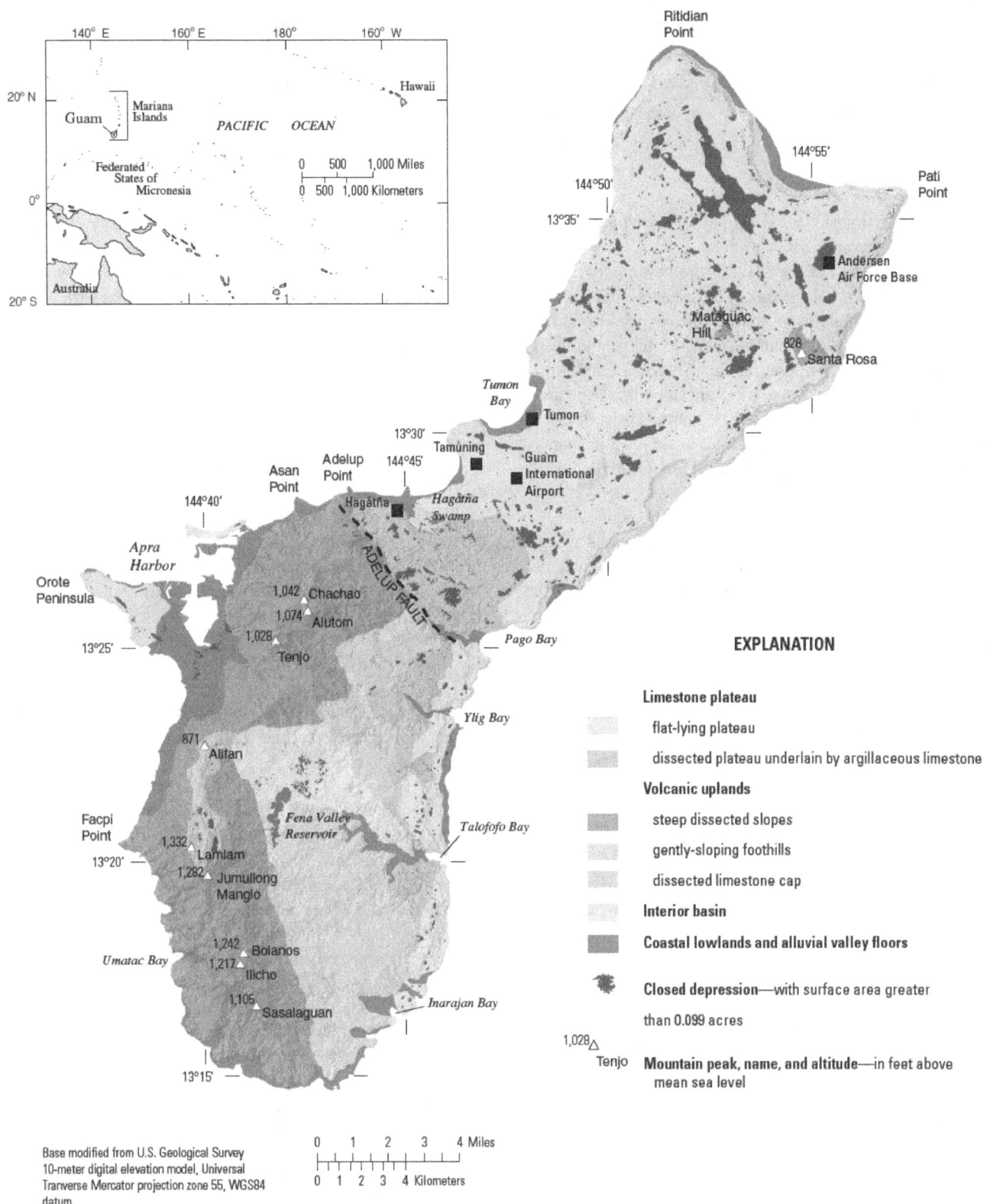

Figure 1. Location map and major physiographic areas of Guam (modified from Tracey and others, 1964; Young, 1988).

when typhoons pass nearby or directly over the island (Lander and Guard, 2003). Rainfall during a year with El Niño conditions tends to be above average (Guard and others, 1999). Some of the driest years occurred during the year following an El Niño event (Lander, 1994).

Hydrogeology

The island of Guam was constructed from a series of water-laid volcanic deposits, upon and around which limestone deposited (Tracey and others, 1964). Volcanic rock forms the foundation of the island and is exposed over 35 percent of the island's surface, predominantly in southern Guam (fig. 3). Limestone overlies the volcanic rock and is exposed over about 60 percent of the island, mainly in northern Guam. The high porosity of the limestone gives it high permeability, whereas the texture and poor sorting of the volcanic rock usually gives it much lower permeability (Gingerich, 2003).

Seven major geologic units are shown in figure 3. The Facpi and Alutom Formations underlie all other exposed rock units (Tracey and others, 1964; Reagan and Meijer, 1984). These formations form much of the volcanic uplands of southern Guam. The Alutom Formation and sediments also crop out in the north at Mount Santa Rosa and Mataguac Hill (see fig. 1). The Facpi Formation consists of lava flows, pillow basalt, and tuffaceous beds of shale and sandstone (Tracey and others, 1964). The Alutom Formation consists of waterlaid lava flows interbedded with pyroclastic rocks ranging from tuffaceous shale to course boulder conglomerate and breccia (Tracey and others, 1964). The permeability of these formations is low (Ward and others, 1965). The Umatac Formation crops out principally in the south-central volcanic uplands and includes reef and forereef limestone, tuff breccias and volcanic conglomerate, pillow lavas, and basalt lava flows (Meijer and others, 1983; Reagan and Meijer, 1984). The permeability of the Umatac Formation is considered low (Ward and others, 1965). Three limestone units (Bonya and Alifan Limestones, and Janum Formation) are grouped together as older limestone in this report. These units overlie the Alutom and Umatac Formations in the southern half of the island, consist of detrital and tuffaceous limestone, and have low to high permeability (Tracey and others, 1964; Ward and others, 1965).

The Barrigada Limestone covers less than 10 percent of the island's surface, but forms the bulk of the aquifer underlying northern Guam (Gingerich, 2003). This formation consists of fine- to course-grained foraminiferal limestone, has high permeability, and supplies water to most of the wells in northern Guam (Tracey and others, 1964; Ward and others, 1965; Contractor and Srivastava, 1990; Jocson and others, 2002). The Mariana Limestone covers most of northern Guam, Orote Peninsula, and the plateau fringing the east coast of southern Guam. This formation consists of reef and lagoonal limestone containing a wide range of lithologies (Tracey and others, 1964). Near the volcanic uplands, the Mariana Limestone is clay-rich (Hagåtña Argillaceous Member), and has moderate to high permeability. The permeability of the non-argillaceous

limestone is considered very high owing to the presence of many solution fissures and channels (Ward and others, 1965; Contractor and Srivastava, 1990; Jocson and others, 2002). Unconsolidated alluvial deposits underlie valley floors and coastal lowlands. The permeability of these formations ranges from low to high (Ward and others, 1965).

On Guam, fresh groundwater occurs in freshwater-lens and perched groundwater systems. A freshwater-lens system consists of a freshwater lens underlain by saltwater. Between this freshwater lens and the underlying saltwater is a zone of mixing containing brackish water. Freshwater-lens systems are found in limestone and volcanic rocks on the island, but the most important sources of groundwater are from the freshwater parts of these systems in the high-permeability limestone rocks of northern Guam (Gingerich, 2003). Perched groundwater is found in areas where low-permeability rocks impede the downward movement of groundwater sufficiently to allow a perched water body to develop above the lowest water table.

The northern half of Guam, which consists of highly permeable limestone, is a groundwater province. No perennial streams flow from the northern limestone plateau to the ocean (Ward and others, 1965). Most rainfall that infiltrates into the land surface to depths below the root zone recharges the freshwater-lens system. In the most permeable limestone, the height of the lowest groundwater table ranges from several feet above sea level in the interior of the island to near sea level at the shore (Gingerich, 2003). Groundwater discharges from the freshwater-lens system as diffuse seepage near the coastline and to subaerial and submarine springs (Taborosi, 2006). In the volcanic rocks at Mataguac Hill, a small amount of perched groundwater discharges at Mataguac Spring.

Southern Guam, which mostly consists of low-permeability volcanic rock, is chiefly a surface-water province. Because of the low permeability of the volcanic rocks, infiltration of rainfall is slow, and large amounts of runoff occur, feeding more than 40 streams. Although the volcanic rock is likely saturated with freshwater at elevations of several hundred feet above sea level in the volcanic uplands, its permeability is too low to yield appreciable water to wells (Ward and others, 1965). Groundwater discharges in stream valleys above sea level where the ground surface intersects the water table. The older limestone units that cap the volcanic uplands contain perched groundwater (Mink, 1976). Springs, such as Almagosa Springs and Dobo Spring, discharge from the perched groundwater systems near the contact between the older limestone units and volcanic rocks.

Soils

The properties and spatial extent of soils on Guam were documented by Young (1988) and updated by the Natural Resources Conservation Service (U.S. Department of Agriculture, 2009). Soil permeability varies according to physiographic region. Most of the soils covering the limestone plateaus and bordering coastal plains have moderately rapid to rapid permeability of 2 to 20 inches per hour (in/h). Soils

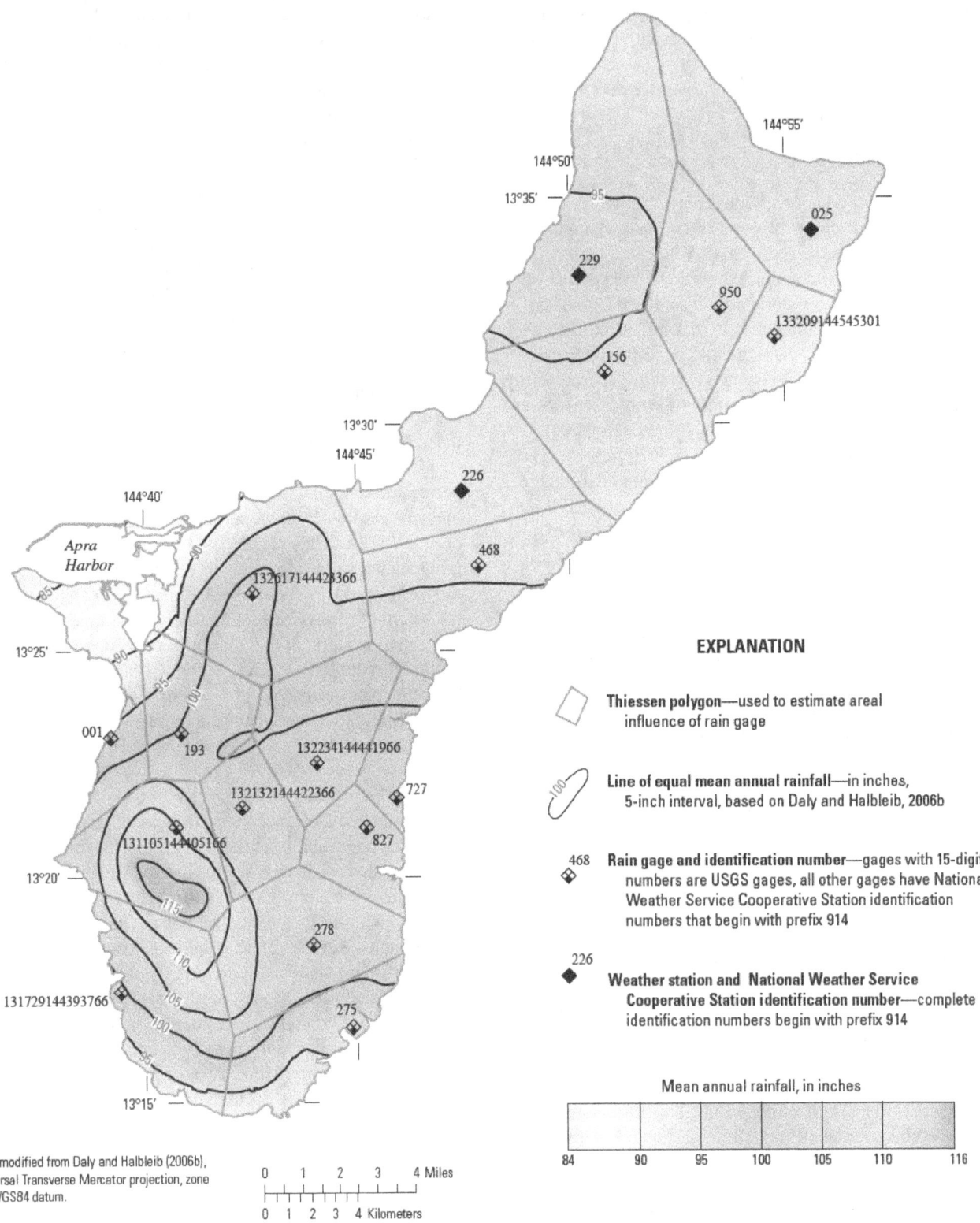

Figure 2. Mean annual rainfall and locations of rain gages and weather stations used in the water-budget calculation for Guam.

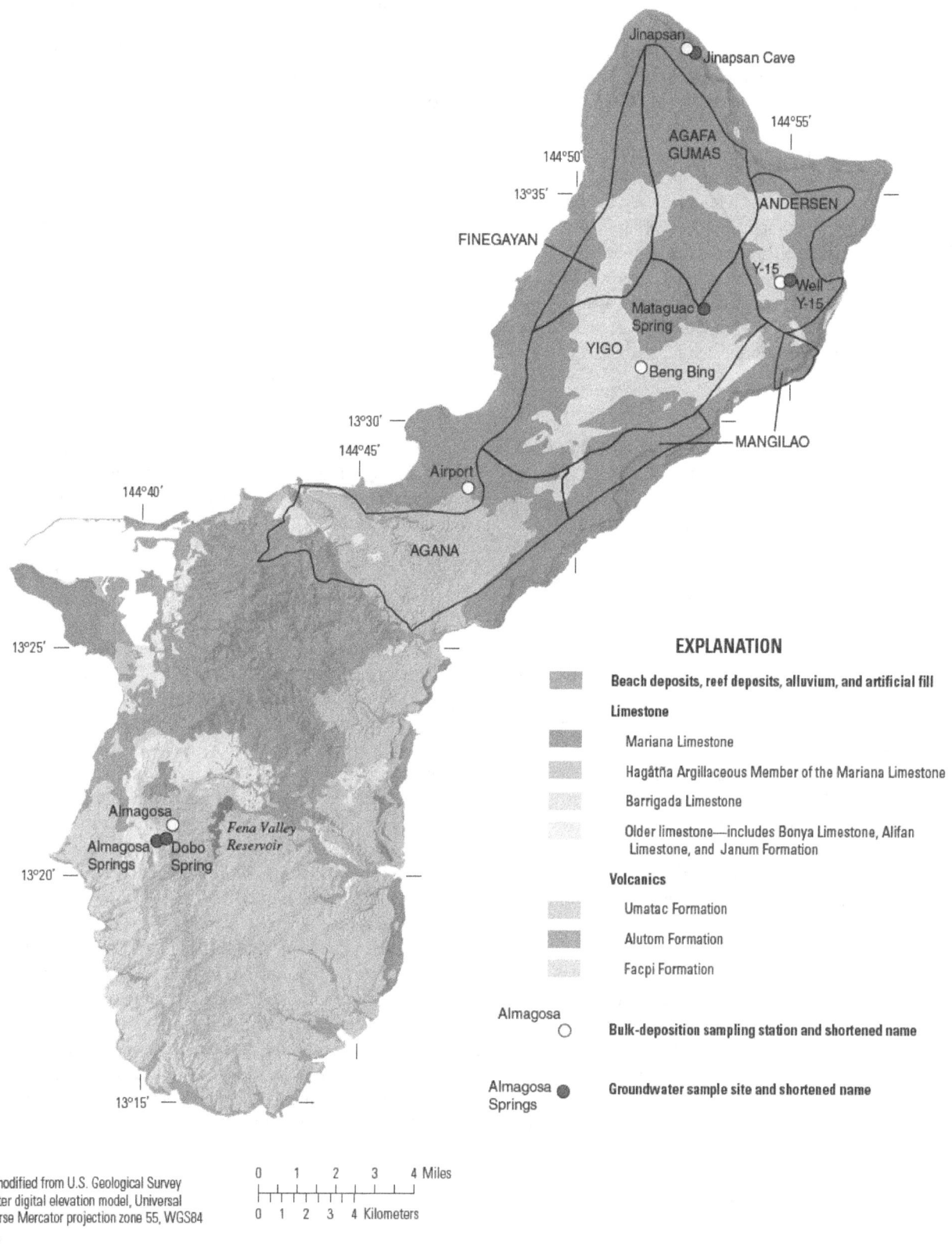

EXPLANATION

| | Beach deposits, reef deposits, alluvium, and artificial fill |

Limestone

	Mariana Limestone
	Hagåtña Argillaceous Member of the Mariana Limestone
	Barrigada Limestone
	Older limestone—includes Bonya Limestone, Alifan Limestone, and Janum Formation

Volcanics

	Umatac Formation
	Alutom Formation
	Facpi Formation

Almagosa ○   Bulk-deposition sampling station and shortened name

Almagosa Springs ●   Groundwater sample site and shortened name

Base modified from U.S. Geological Survey 10-meter digital elevation model, Universal Tranverse Mercator projection zone 55, WGS84 datum.

0 1 2 3 4 Miles

0 1 2 3 4 Kilometers

Figure 3. Geologic map of Guam (modified from Tracey and others, 1964; Siegrist and others, 2008), bulk-deposition sampling stations, and groundwater sampling sites. Boundaries of named aquifer subbasins, as defined by Camp, Dresser & McKee Inc. (1982), are shown in black.

covering the argillaceous limestone areas are relatively less permeable because of their higher clay content. Volcanic uplands are covered by soils that generally have moderately slow to moderate (0.2 to 2 in/h) permeability. Older limestone in the southern highlands is covered by soils with moderately rapid to rapid permeability. Soils covering the coastal lowlands and valley floors that are derived from volcanic alluvium have slow to moderate permeability (0.01 to 2 in/h).

Soil thickness across Guam tends to vary according to parent material and topography. Soils covering much of the northern limestone plateau are generally no more than 10 in. thick, except in closed depressions. Soils covering the argillaceous limestone soils and volcanic uplands are spatially variable but are typically thicker than 10 in. Soils in closed depressions, coastal lowlands, or valley floors may be several feet thick.

Land Cover

According to the U.S. Department of Agriculture (2006a) vegetation map (fig. 4), about half of Guam's land surface is covered by forests; most other areas are urbanized or are covered by grasslands. Owing to a long history of disturbances, (typhoons, World War II, and human development) little undisturbed primary forest exists on the island (Fosberg, 1960). Consequently, the forests on Guam are mostly dense, tangled thickets of weedy, secondary trees that surround scattered trees of taller stature (Fosberg, 1960). Most trees are between 15 and 30 ft tall and have a diameter less than 10 in. (Donnegan and others, 2004). Forest land-cover categories "Limestone forest" and "Scrub forest" cover much of the undeveloped areas underlain by limestone. Disturbed areas within these forests are often covered by dense thickets of *Leucaena leucocephala* (Fosberg, 1960). *L. leucocephala,* known to Guamanians as tangantangan, is estimated to be the most abundant species of tree on Guam (Donnegan and others, 2004). Continuous patches of *L. leucocephala* are mapped as "*Leucaena* stand." Urbanized areas are classified as "Urban builtup" and "Urban cultivated" and are predominantly in the northern half of the island. In the southern half of the island, the volcanic uplands are mostly covered by "Savanna complex" or "Ravine forest." Savanna complex is a mixed grassland dominated by *Miscanthus floridulus* (sword grass), which grows in dense clumps up to 10 ft high (Fosberg, 1960). Ravine forest consists of trees, generally brushy and low in stature, which form a dense and irregular canopy (Fosberg, 1960). All other land cover classes total less than 7 percent of the land surface.

Estimates of Groundwater Recharge

Water-Budget Model

The water-budget model is designed to simulate—on a daily basis—the hydrological processes and physical conditions that affect recharge on Guam. Hydrological processes included here are rainfall, irrigation, septic-system leachate, water-main leakage, runoff, and evapotranspiration. Physical parameters are land cover (vegetation), geology, and moisture-storage capacity of the soils.

Conceptual Model

Groundwater recharge for Guam is estimated using a daily water-budget model that is a "threshold-type" or "reservoir" model based on the Thornthwaite and Mather (1955) mass-balance procedure. The structure of the model is similar to previous water-budget models developed by the U.S. Geological Survey (USGS) for recharge studies in the Hawaiian Islands (Oki, 2002; Izuka and others, 2005; Engott and Vana, 2007). Following the method developed by Engott (2011) for the island of Hawai'i, the model developed for this study also accounts for canopy evaporation in forests. Unlike the model used by Engott (2011), however, the method used in this study to estimate canopy evaporation is a function of daily rainfall and the canopy storage capacity of a forest. Four generalized flow diagrams are used in this study—one for forest land covers, one for nonforest land covers without impervious surfaces (fig. 5), one for urbanized subareas without storm-drain systems, and one for urbanized subareas with storm-drain systems (fig. 6). All conceptual models employ a plant-root zone reservoir. The forest model includes a second reservoir consisting of the forest canopy.

The volume of the plant-root zone reservoir is based on plant and soil properties. The model accounts for water entering, leaving, and being stored within the plant-root zone reservoir on a daily basis. At the end of a given day, if the volume of water entering the system exceeds the storage capacity of the plant-root zone reservoir, given the antecedent water content and water losses due to evapotranspiration processes, the reservoir overflows. This overflow is counted as groundwater recharge by the model.

The forest-canopy reservoir is not treated as a true reservoir in the model calculations, as the volume of water within the reservoir is not tracked. Instead, net precipitation is output from the reservoir that becomes input to the plant-root zone. Net precipitation is calculated using a relation to daily rainfall and the canopy storage capacity of a forest that was derived from Wallace and McJannet (2006) and was modified for this model using results from published net precipitation studies. Canopy evaporation is then calculated as the difference between rainfall volume and net precipitation.

Model Exclusions and Assumptions

Several exclusions and assumptions are made to simplify the water-budget model. Groundwater recharge is water that has infiltrated into the land surface and percolated below the root zone. Because the objective of this study is to estimate long-term mean groundwater recharge, the duration it takes water to percolate through the vadose zone to the lowest water

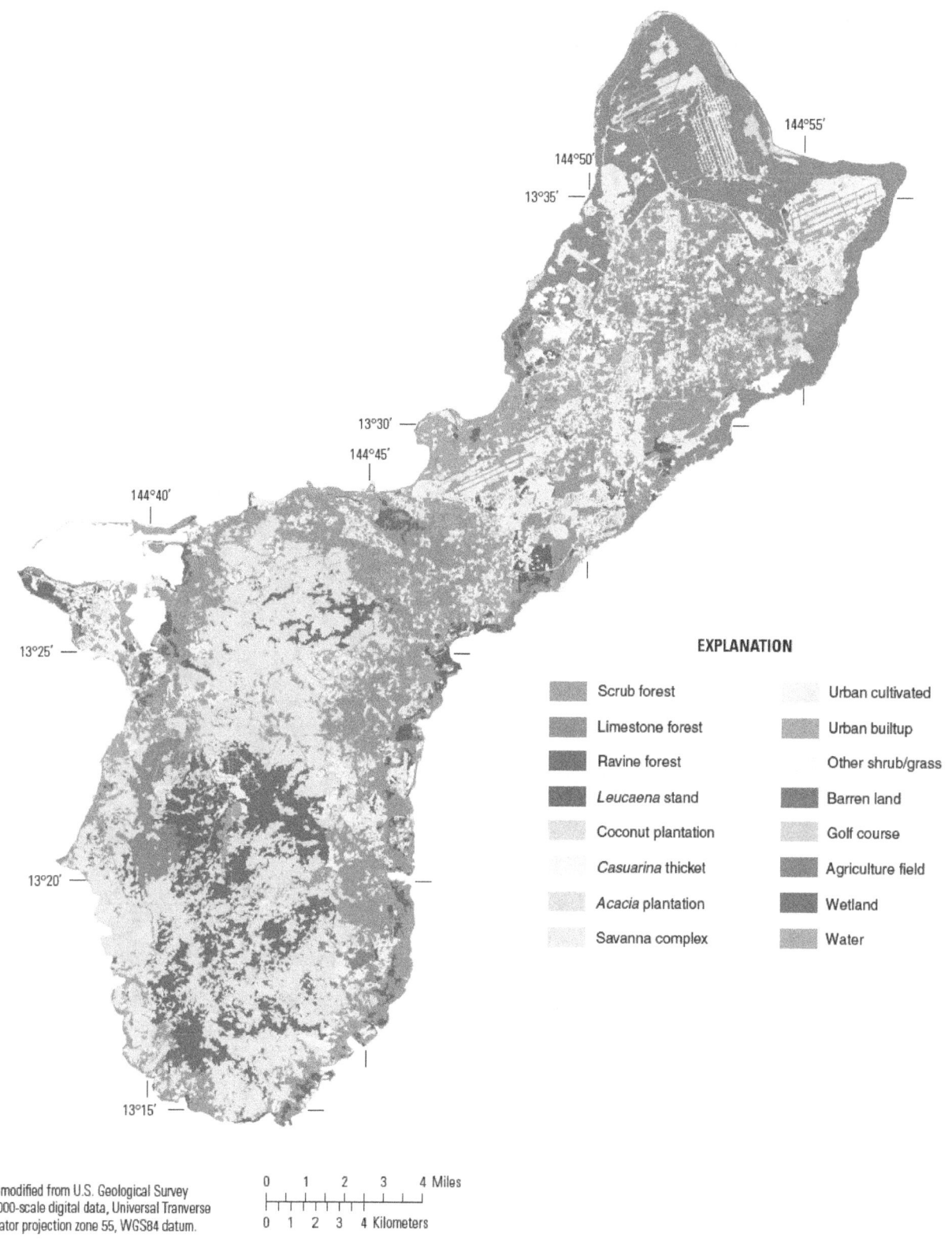

Figure 4. Land cover on Guam (modified from U.S. Department of Agriculture, 2006a). *Acacia* plantation, barren land, and *Casuarina* thicket land covers are difficult to see at this scale.

FOR FOREST LAND COVERS:
(modified from McJannet and others, 2007)

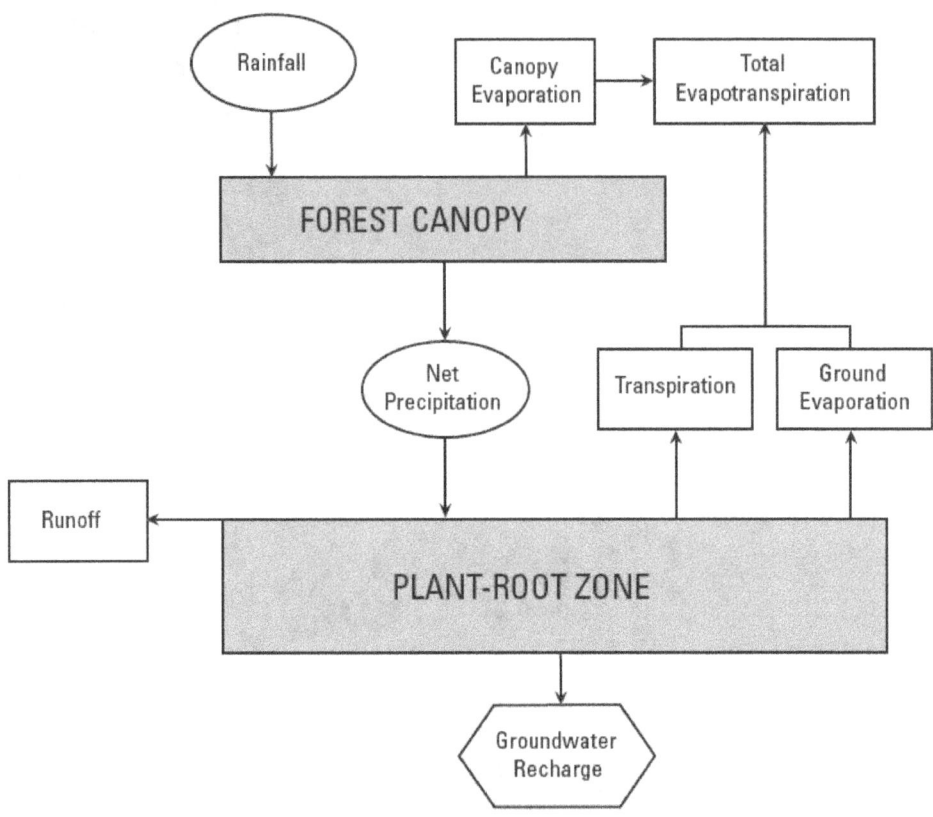

FOR NONFOREST LAND COVERS
WITHOUT IMPERVIOUS SURFACES:

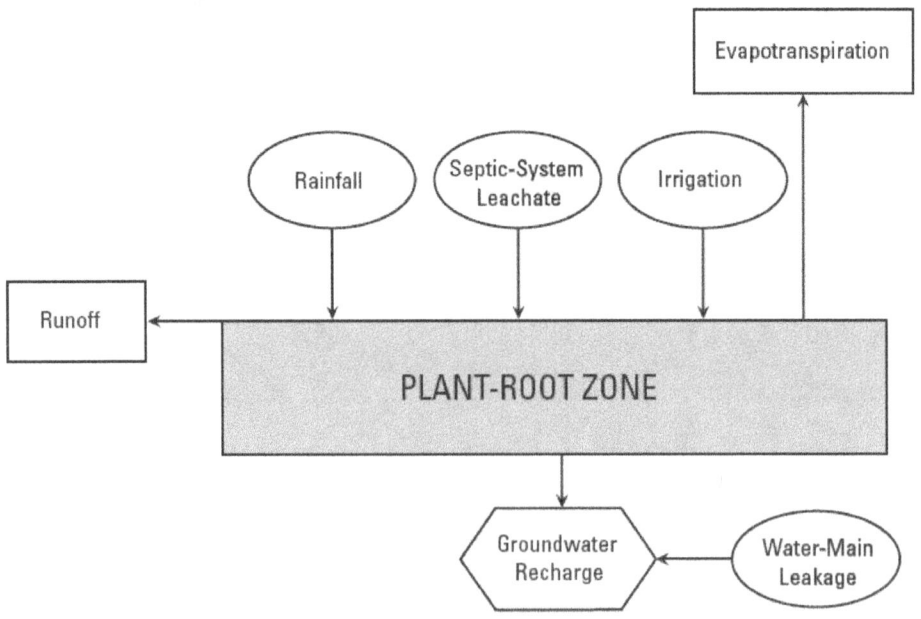

Figure 5. Generalized water-budget flow diagrams for forest and nonforest land covers without impervious surfaces (modified from Engott, 2011).

FOR URBANIZED SUBAREAS WITHOUT STORM-DRAIN SYSTEMS

W = excess water from impervious fraction added to pervious fraction of subarea

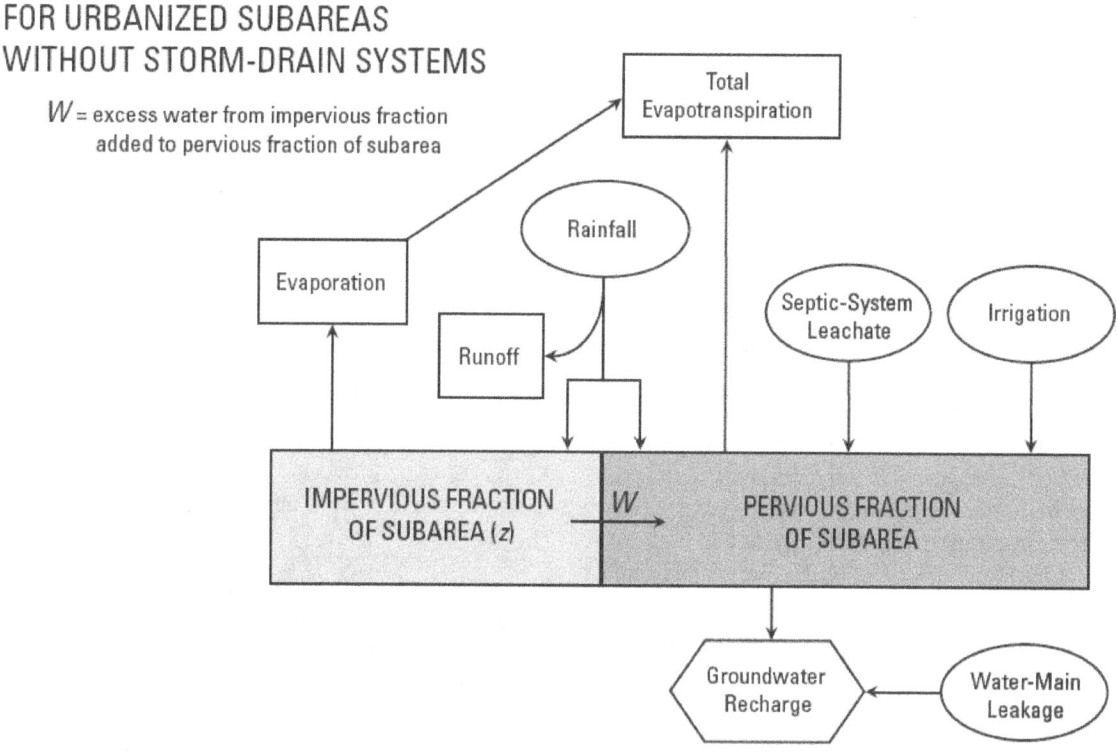

FOR URBANIZED SUBAREAS WITH STORM-DRAIN SYSTEMS

W = excess water from impervious fraction added to pervious fraction of subarea

SD = excess water from impervious fraction captured by storm-drain systems

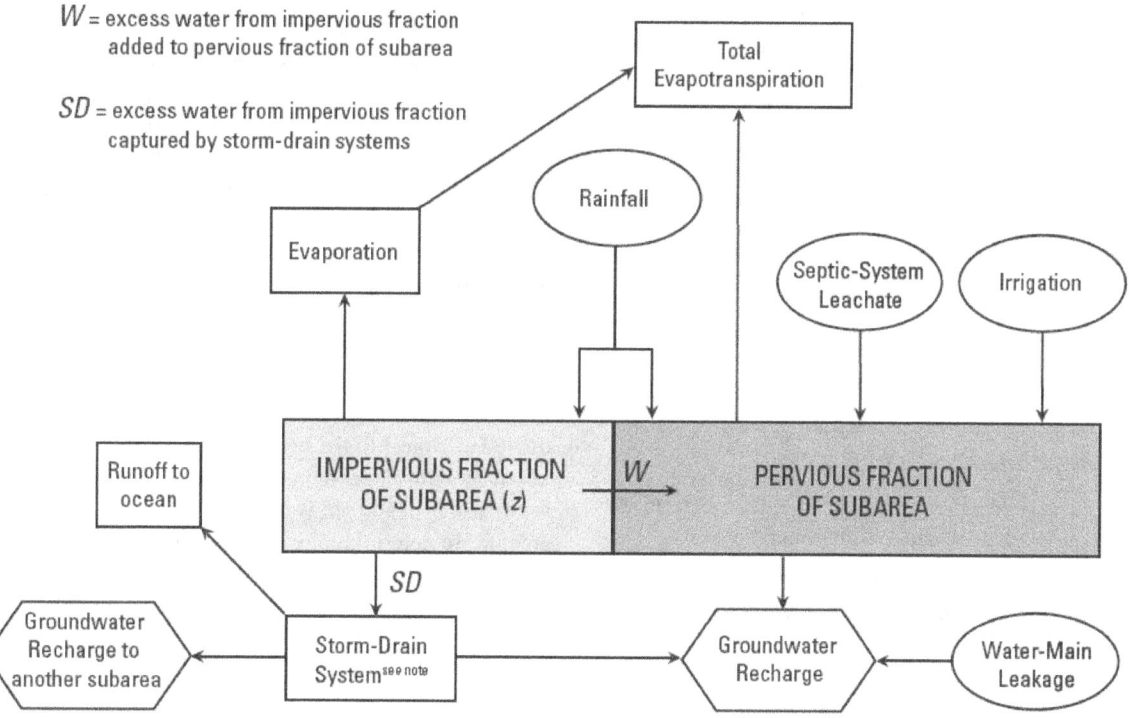

Note: flow out of storm-drain system depends on which storm-drain zone the subarea is located in.

Figure 6. Generalized water-budget flow diagrams for urbanized subareas without and with storm-drain systems. Urbanized subareas include the land-cover categories urban cultivated and urban builtup.

table is not considered. Recharge from streambed seepage and the variability of soil moisture with regard to depth within the soil root zone are not considered. Water input from fog interception is assumed to be negligible because the climate conditions characteristic of cloud forests are not prevalent on Guam (Fosberg, 1960).

Model Calculations

Groundwater recharge for Guam was computed using the daily water-budget model and input data that quantify the spatial and temporal distributions of rainfall, irrigation, water-main leakage, septic-system leachate, evapotranspiration, runoff, soil type, and land cover. Areas of homogeneous properties, termed "subareas," are generated by merging datasets that characterize the spatial and temporal distributions of rainfall, irrigation, water-main leakage, septic-system leachate, reference evapotranspiration, runoff, soil type, and land cover in a geographic information system (GIS). For each subarea, recharge is calculated by the water-budget model. At the end of a simulation period, results for the subareas are summed over larger areas of interest, which can include entire aquifer sectors. The water-budget model for Guam has 138,408 subareas, with an average area of 0.97 acres.

For each subarea at the start of each day, the model calculates an interim moisture storage. Interim moisture storage is the amount of water that enters the plant-root zone for the current day plus the amount of water already in the zone from the previous day. For the first day of the simulation, a value for the amount of water already in the zone from the previous day (initial soil moisture) is selected by the user. For nonforest subareas, interim moisture storage, expressed in units of length [L], is given by the equation

$$X_i = P_i + F_i + I_i + W_i - R_i + S_{i-1}, \qquad (1a)$$

where

X_i = interim moisture storage for current day [L],

P_i = rainfall for current day [L],

F_i = septic-system leachate [L],

I_i = irrigation for current day [L],

W_i = excess water from the impervious fraction of an urban area distributed over the pervious fraction [L],

R_i = runoff for current day [L],

S_{i-1} = moisture storage at the end of the previous day (i-1) [L], and

i = subscript designating current day.

For forest subareas, interim moisture storage is given by the equation

$$X_i = (NP)_i - R_i + S_{i-1}, \qquad (1b)$$

where

$(NP)_i$ = net precipitation for current day [L],

For forest subareas, net precipitation is computed as rainfall minus canopy evaporation, which is the amount of water from rainfall that collects on the leaves, stems, and trunks of trees and subsequently evaporates. The equation is

$$(NP)_i = P_i - (CE)_i, \qquad (2)$$

where

$(CE)_i$ = canopy evaporation [L].

For urbanized subareas, the interim equation includes the factor W_i, which is a function of the fraction of urban subareas that are estimated to be impervious (see equation 1a). Urbanized subareas, which include the land-cover categories urban cultivated and urban builtup, are assigned a fraction (z) that is impervious. This fraction is used to separate, from the total rain that falls in an urbanized subarea, a depth of water that is treated computationally as though it fell on an impervious surface (fig. 6). Of the rain that falls on the impervious fraction of the subarea, some water is subtracted to account for direct evaporation, and the remaining water (U_i) is the excess water from the impervious fraction of the subarea. For urbanized areas that are not within drainage areas of storm-drain systems, all excess water from the impervious fraction of the subarea is added to the water budget of the pervious fraction of the model subarea, and W_i is equal to U_i. For urbanized areas within drainage areas of known storm-drain systems, a fraction of U_i is assumed to be captured by the storm-drain system, and the remainder is added to the pervious fraction of the model subarea. Depending on the storm-drain system, the fraction of excess water captured by the system (SD_i) for a given subarea is either added as direct recharge to the given subarea, added as direct recharge to another subarea of the model, or subtracted as runoff to the ocean.

For an urbanized model subarea, excess water, U_i, and water storage (ponded water) on the surface of impervious areas were determined using the following conditions:

$$X1_i = P_i - R_i + T_{i-1}, \qquad (3)$$

for $X1_i \leq N$, $U_i = 0$, and $X2_i = X1_i$,

for $X1_i > N$, and $z \neq 1$, $U_i = (X1_i - N)z / (1-z)$,

and $X2_i = N$,   (4)

where

$X1_i$ = first interim moisture storage on the surface of impervious area for current day [L],

$X2_i$ = second interim moisture storage on the surface of impervious area for current day [L],

T_{i-1} = water storage (ponded water) on the surface of impervious area at the end of the previous day (i-1) [L],

N = rainfall interception capacity (maximum amount of water storage on the surface of impervious area) [L], and

z = fraction of area that is impervious [dimensionless].

Equations 3 and 4 are only used for urbanized subareas. All urbanized subareas are assigned z-values that are greater than 0 and less than 1.

The fraction of the excess water from the impervious fraction that is added to the pervious fraction of the subarea is determined from

$$W_i = U_i - SD_i, \qquad (5)$$

where

SD_i = excess water captured by a storm-drain system [L].

For subareas within the drainage area of a storm-drain system, the excess water captured by a storm-drain system is determined from

$$SD_i = y \times U_i, \qquad (6)$$

where

y = fraction of excess water captured by a storm-drain system [dimensionless].

The water storage on the surface of the impervious area at the end of the current day, T_i, is determined from the equation

for $X2_i > V_i$, $\quad T_i = X2_i - V_i$, and

for $X2_i \leq V_i$, $\quad T_i = 0$, $\qquad (7)$

where

V_i = evaporation for current day [L].

The next step in the water-budget computation is to determine the amount of water that will be removed from the plant-root zone by evapotranspiration (ET). Actual ET is a function of potential ET and interim moisture (X_i). A vegetated surface loses water to the atmosphere at the potential-ET rate if sufficient water is available. At all sites, potential ET was assumed to be equal to reference evapotranspiration multiplied by an appropriate vegetation factor, termed a crop coefficient. For moisture contents greater than or equal to a threshold value,

C_i, the rate of ET, expressed in units of length per time [L/T], was assumed to be equal to the potential-ET rate. For moisture contents less than C_i, ET was assumed to occur at a reduced rate that declines linearly with soil-moisture content:

for $S \geq C_i$, $\qquad E = (PE)_i$, and

for $S < C_i$ and $C_i > 0$ $\quad E = S \times (PE)_i / C_i$, $\qquad (8)$

where

E = instantaneous rate of evapotranspiration [L/T],

$(PE)_i$ = potential-evapotranspiration rate for the current day [L/T],

S = instantaneous moisture storage [L], and

C_i = threshold moisture storage for the current day below which evapotranspiration is less than the potential-evapotranspiration rate [L].

The threshold moisture storage, C_i, was estimated using the model of Allen and others (1998) for soil moisture. In this model, a depletion fraction, p, which ranges from 0 to 1, is defined as the fraction of maximum moisture storage that can be depleted from the root zone before moisture stress causes a reduction in ET. The threshold moisture, C_i, is estimated from p by the equation

$$C_i = (1 - p) \times S_m, \qquad (9)$$

where

S_m = moisture-storage capacity of the plant-root zone [L].

The moisture-storage capacity of the plant-root zone, S_m, expressed as a depth of water, is equal to the plant root depth multiplied by the available water capacity of the soil, ϕ. Available water capacity is the difference between the volumetric field-capacity moisture content and the volumetric wilting-point moisture content:

$$S_m = D \times \phi, \qquad (10)$$

where

D = plant root depth [L],

ϕ = $\theta_{fc} - \theta_{wp}$ [L³/L³],

θ_{fc} = volumetric field-capacity moisture content [L³/L³], and

θ_{wp} = volumetric wilting-point moisture content [L³/L³].

Values for p depend on vegetation type and can be adjusted to reflect different potential-ET rates. In this study, p values were based on data in Allen and others (1998).

In the water-budget model, the ET rate from the plant-root zone may be (1) equal to the potential-ET rate for part of the day and less than the potential-ET rate for the remainder of the day, (2) equal to the potential-ET rate for the entire day, or (3) less than the potential-ET rate for the entire day. The total ET from the plant-root zone during a day is a function of the potential-ET rate $((PE)_i)$, interim moisture storage (X_i), and threshold moisture content (C_i). By recognizing that $E = -dS/dt$, the total depth of water removed by ET during a day, E_i, was determined as follows:

for $X_i > C_i$ and $C_i > 0$,
$$E_i = (PE)_i t_i + C_i\{1-\exp[-(PE)_i(1-t_i)/C_i]\},$$
for $X_i > C_i$ and $C_i = 0$,
$$E_i = (PE)_i t_i,$$
for $X_i \leq C_i$ and $C_i > 0$,
$$E_i = X_i\{1-\exp[-(PE)_i / C_i]\},$$

and

for $X_i = C_i$, and $C_i = 0$,
$$E_i = 0, \tag{11}$$

where

E_i = evapotranspiration from the plant-root zone during the day [L],

t_i = time during which moisture storage is above C_i [T]. It ranges from 0 to 1 day and is computed as follows:

for $(X_i - C_i) < (PE)_i(1 \text{ day})$
$$t_i = (X_i - C_i)/(PE)_i,$$

and

for $(X_i - C_i) \geq (PE)_i(1 \text{ day})$,
$$t_i = 1. \tag{12}$$

After accounting for runoff (equation 1a or 1b), ET from the plant-root zone for a given day was subtracted from the interim moisture storage, and any moisture remaining above the maximum moisture storage was assumed to be recharge. The daily rate of direct recharge is also added at this point. Direct recharge includes water-main leakage, which is assumed to occur at depths below the root zone and is not subject to ET. Direct recharge also includes stormwater runoff that is routed into drywells or the Harmon Sink. Recharge and moisture storage at the end of a given day were assigned according to the following conditions:

for $X_i\text{-}E_i \leq S_m$, $Q_i = DR$, and $S_i = X_i\text{-}E_i$,

and

for $X_i\text{-}E_i > S_m$, $Q_i = (X_i\text{-}E_i\text{-}S_m) + DR$,

and $S_i = S_m$, $\tag{13}$

where

Q_i = groundwater recharge during the day [L], and
S_i = moisture storage at the end of the current day (i) [L], and
DR = daily rate of direct recharge [L] (a constant).

For urbanized subareas, recharge is prorated to the entire subarea by multiplying the recharge depth from the pervious fraction by the pervious fraction of the subarea. Moisture storage at the end of the current day, expressed as a depth of water, is equal to the root depth multiplied by the difference between the volumetric soil-moisture content within the root zone at the end of the current day and the volumetric wilting-point moisture content:

$$S_i = D \times (\theta_i - \theta_{wp}), \tag{14}$$

where

θ_i = volumetric soil-moisture content at the end of the current day, i, $[L^3/L^3]$.

Fast Flow

On Guam, during and immediately after heavy rainfall, water may rapidly infiltrate into the ground in certain limestone areas through preferred pathways bypassing the soil layer and bedrock matrix (Jocson and others, 2002). This phenomenon is defined as "vadose fast flow" by Jocson and others (2002). Contractor and Jenson (2000) included a parameter (SINK) within their groundwater model of northern Guam that accounts for vadose fast flow. A SINK value of 33 percent of rainfall yielded their best fit between observed and modeled water levels at four observation wells in the Yigo aquifer subbasin.

In this report, to simulate vadose fast flow, a second water-budget model was developed. This second water-budget model (method 2) is nearly identical to the water-budget model described above (method 1), differing only in the order in which evapotranspiration is accounted for. For method 1, evapotranspiration is first subtracted from the interim soil-moisture storage, and any soil moisture remaining above the soil-moisture storage capacity is assumed to be recharge. For method 2, any interim soil moisture greater than the soil-moisture capacity is assumed to be recharge, and evapotranspiration is then subtracted from the remaining soil-moisture storage. Method 1 is more representative of what Jocson and others (2002) refer to as "vadose percolation," whereas method 2 is more representative of vadose fast flow. Because Contractor and Jenson (2000) estimated a SINK value of 33 percent of rainfall for their groundwater model, recharge for all areas underlain by limestone was computed as 33 percent of recharge calculated using method 2 plus 67 percent of recharge calculated using method 1. For areas not underlain by limestone, recharge was calculated using only method 1.

Table 2. Land-cover parameters used in water-budget calculations for Guam.

[Crop-coefficient values for forest land covers are used to compute the combination of transpiration and ground evaporation; canopy evaporation is calculated separately; <, less than]

Land-cover description	Fraction of total land area	Pervious fraction	Crop coefficient	Root depth (inches)	Depletion fraction
Forest land covers					
Scrub forest	0.23	1	0.69	20	0.50
Limestone forest	0.13	1	0.69	20	0.50
Ravine forest	0.08	1	0.57	20	0.50
Leucaena stand	0.03	1	1.04	20	0.50
Coconut plantation	<0.01	1	0.48	20	0.50
Casuarina thicket	<0.01	1	0.62	20	0.50
Acacia plantation	<0.01	1	0.62	20	0.50
Nonforest land covers					
Savanna complex	0.21	1	1.23	24	0.60
Urban cultivated	0.13	0.82	0.86	12	0.40
Urban builtup	0.13	0.40	1.22	12	0.50
Other shrub/grass	0.02	1	1.00	20	0.50
Barren land	<0.01	1	1.15	5	0.60
Golf course	<0.01	1	0.86	30	0.40
Agriculture field	<0.01	1	0.96	12	0.45
Wetland	<0.01	1	1.07	6	0.50
Water	<0.01	1	1.05	0	1.00

Model Input

Land-Cover Map

The U.S. Department of Agriculture (2006a) vegetation map was used as the base land-cover map. Modifications to the land-cover map were performed in a GIS. Modifications included combining land-cover classes and classifying golf-course areas. Similar land-cover classes were reclassified into a single land-cover class in order to simplify the land-cover map. Bad land, barren, and sand beach/bare rocks classes were reclassified as barren. Mangrove swamp, marshland, and wet land classes were reclassified as wetland. Strand vegetation was incorporated into the other shrub/grass class. Golf courses were delineated using a high resolution orthoimage (U.S Department of Agriculture, 2006b) and were inserted onto the land-cover map. The modified land-cover map used for the baseline scenarios is termed "2004 land cover" throughout the remainder of this report because it was derived from 2004 satellite images.

Urbanized areas on Guam are classified as "urban builtup" or "urban cultivated" in the 2004 land-cover map. Urban builtup includes areas that are mostly buildings, roads, or other paved areas. Urban cultivated includes maintained grassy areas and other vegetation around cities and military areas. For these land covers, the fraction of area that is pervious was estimated in GIS using a map of impervious surfaces (National Oceanic and Atmospheric Administration, 2009). All other land covers were assigned a pervious fraction of 1 (table 2).

Rainfall

Mean Rainfall

Mean monthly rainfall maps created by Daly and Halbleib (2006b) were used as the basis for the spatial distribution of daily rainfall. Using GIS, each monthly map was converted from raster grids to polygon bands with 0.1-inch rainfall increments. The mean monthly rainfall maps are representative of the period 1971–2000. Mean annual rainfall distribution across Guam was produced by summing the mean monthly rainfall grids (Daly and Halbleib, 2006b) (fig. 2).

Records from 18 rain gages on Guam were used to generate daily rainfall values for the model. Thiessen polygons were used to spatially apply the daily rainfall patterns indicated by the gages (fig. 2). Rain gages were selected on the basis of completeness of daily record and location. Rainfall data were obtained from the National Climatic Data Center (NCDC) and the USGS. In this report, all weather stations and rain gages that were not operated by the USGS are referred to by their National Weather Service Cooperative Station Network identification number. Rain gages operated by the USGS are referred to by their 15-digit USGS station identification number.

For input to the water-budget model, daily rainfall was synthesized by disaggregating the monthly rainfall values using the method of fragments (see, for example, Oki, 2002). Daily rainfall sequences were synthesized because each rain gage used in the analysis had missing or incomplete daily rainfall records for the long-term period of interest used in the water-budget model.

The synthesized daily data approximate the long-term average character of daily rainfall, such as frequency, duration, and intensity, but do not reproduce the actual historical daily rainfall record. The method of fragments creates a synthetic sequence of daily rainfall from monthly data by imposing the rainfall pattern from a rain gage with daily data. Fragments were created by dividing each daily rainfall measurement for a particular month by the total rainfall for that month. This created a set of fragments for that particular month in which the total number of fragments was equal to the number of days in the month. Fragment sets were created for every gage for every month in which complete daily rainfall measurements were available. Fragment sets were grouped by month of the year and by rain gage. The fragment set to be used for a given gage for a given month was selected randomly from among all available sets for that gage for that month of the year. Synthesized daily rainfall for a given month was created by multiplying total rainfall for that month by each fragment in the set, thereby providing daily rainfall, P_i, for equation 1a or 1b.

Temporal Variability

Variations in annual rainfall were accounted for to provide a more realistic estimation of long-term mean groundwater recharge. The magnitude in annual variability across the island was based on observed rainfall at weather station 914226, located at the Guam International Airport. This station was selected on the basis of its completeness of record.

The water-budget model was run for a period of 45 years to account for annual variations in rainfall that occurred during 1961–2005 (fig. 7). This period was selected on the basis of the completeness of rainfall records at weather station 914226 and the availability of data from which to calculate potential evaporation. For the few years with incomplete records, estimates listed in Lander and Guard (2003) were used. Observed annual rainfall at weather station 914226 for each year during the 45-year period was divided by the mean annual rainfall for 1971–2000, creating a series of 45 weighting factors. Rainfall for a given month was calculated by multiplying the mean monthly value, derived from Daly and Halbleib (2006b), by the weighting factor appropriate for the year. This assumes the monthly rainfall distribution, based on maps from Daly and Halbleib (2006b) and specific to 1971–2000, was the same during 1961–2005.

Irrigation

Daily irrigation was applied to the agricultural-field and golf-course land covers in the water-budget model. To be conservative, irrigation was not applied to subareas classified as urban cultivated because the percentage of these subareas that are actually irrigated is not known, but is most likely less than 100 percent. The total area of agriculture-field and golf-course land covers is less than 2 percent of the island. Hence, the effect of irrigation on the regional water budget

is small. Nonetheless, irrigation does affect the soil-moisture content and recharge, so it is accounted for here. Because records of irrigation application rates are not available, irrigation was calculated as the difference between monthly potential ET and rainfall. This is similar to the approach used for recent water budgets in Hawai'i (Engott and Vana, 2007; Engott, 2011). For agricultural areas, irrigation was applied uniformly over each day in the month. Golf-course irrigation was applied in the same manner, but only during the dry season months of January through June, based on discussions with golf-course maintenance personal on Guam. Irrigation is calculated as follows:

$$\text{for } (PE)_m - P_m > 0, \quad I_m = ((PE)_m - P_m) / d_m,$$
and
$$\text{for } (PE)_m - P_m \le 0, \quad I_m = 0, \tag{15}$$

where

I_m = the amount of daily irrigation for month m [L],

P_m = the amount of rainfall for month m [L],

$(PE)_m$ = the potential evapotranspiration for month m (varies by land cover) [L], and

d_m = the number of days in month m [dimensionless].

Septic Systems

Recharge is enhanced where households dispose wastewater into septic systems. On Guam, about 15,000 households use on-site septic systems (Guam Waterworks Authority, 2007a). Thus, water leaching from septic systems is important to include in the Guam water-budget model. Septic system locations and leaching rates were estimated on the basis of surveys and published information.

Locations of households that use septic systems are not completely known at this time, but it is estimated that 42, 44, and 14 percent of these households are located in GWA's designated northern, central, and southern regions, respectively (Guam Waterworks Authority, 2007a). In 2007, the Water and Environmental Research Institute (WERI) of the Western Pacific mapped the locations of about 6,200 households on Guam that have septic systems. This study used GIS data provided by WERI that specified the location and "sewer status" of existing GWA customer accounts. The sewer status of each account was classified as either (1) connected to public sewer system, (2) not connected to public sewer system, or (3) unknown. On the basis of GWA's estimate of 15,000

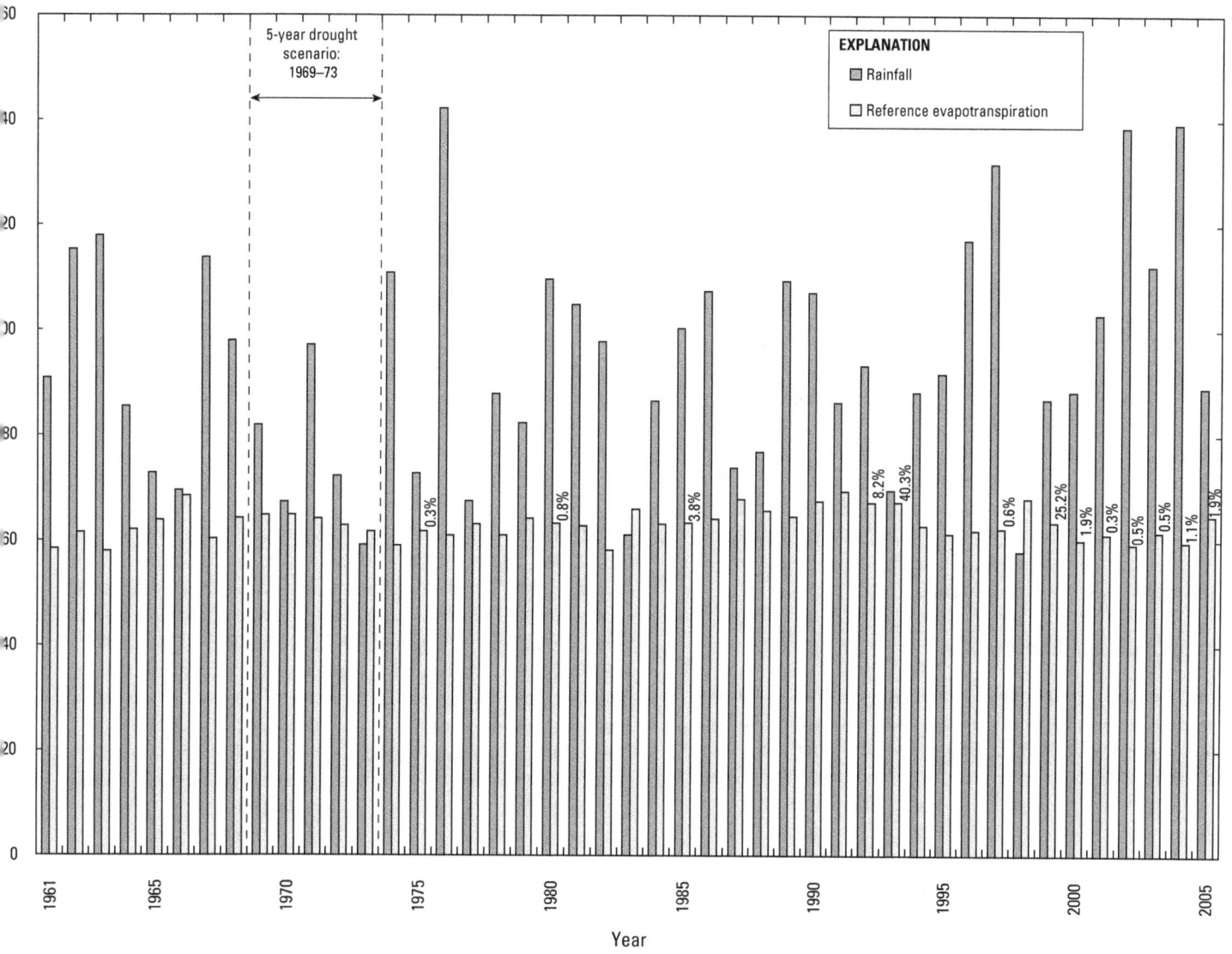

Figure 7. Annual reference evapotranspiration (calculated using the method of Allen and others [1998]), and annual rainfall at [ ] International Airport (National Weather Service station 914226), 1961–2005. Solar radiation data used to calculate reference [ev]apotranspiration were from National Weather Service station 914229 (1961–1990) and station 914226 (1991–2005). The percentage [da]ys in each year for which reference evapotranspiration was calculated using weather observations (air temperature, dew point [temp]erature, and wind speed) that were estimated from weather observations at National Weather Service stations 914229 and 914025 [are s]hown above the annual reference evapotranspiration bars; no numbers are shown above the annual reference evapotranspiration [bars] for years in which all these weather observations from National Weather Service station 914226 were used to calculate reference [evap]otranspiration.

households with septic systems, about 8,800 of the accounts classified with an "unknown" sewer status were selected and assumed to have septic systems. Accounts furthest from sewer mains were selected so that the total number of houses with septic systems within the northern, central, and southern regions agreed with the proportions reported by the Guam Waterworks Authority (2007a). Sewer mains were located on the basis of GIS data provided by GWA.

A septic-system map specifying areas on Guam that receive wastewater from septic systems was created using a GIS. Depending on the type of septic system, wastewater is ultimately disposed into a leach field or seepage pit (cesspool) within the ground. To simulate leach-field and cesspool areas on the septic-system map, a 750 ft^2 circular area was created around each household assumed to have a septic system. The 750 ft^2 area was based on the Guam Environmental Protection Agency (1997) leach field absorption area requirements for a four-bedroom household.

The leaching rate from septic systems was calculated from average household water use. On Guam, average daily water use for a residential account is 339 gallons (Guam Waterworks Authority, 2007b). For this study, it is assumed that 95 percent of water used by a household becomes wastewater. Thus, 322 gallons was applied daily to each area representing a septic-system leach field. Based on the Guam Environmental Protection Agency (1997) leach field depth requirements, this water was added to the plant-soil zone, and was subject to evapotranspiration.

Water-Main Leakage

Leakage from Guam's major water-main systems is thought to be substantial, though exact leak rates and locations are not well known. Because water mains can transport water from water source areas in certain parts of the island to recipients in different parts of the island, it is important to account for water leakage. This study accounts for leakage from the Andersen AFB, Navy, and GWA water-main systems. Water-main leakage is applied along the lengths of each water-main system and is added as direct recharge. It is assumed that the leakage enters the soil below the plant-soil zone (see equation 13) and is not subject to ET.

Leakage rates for each water-main system are based on current estimates of unaccounted for water (UFW). For a given water-main system, UFW equals water supplied to the system minus measured water consumption from a system. Water consumption is measured by meters at recipient locations. UFW can be due to overflows or leakage from the distribution system, or undocumented (unmetered, or inaccurately metered) water consumption. Current UFW rates from Andersen AFB and Navy systems are estimated to be 0.71 and 1.62 million gallons per day (Mgal/d), respectively (Travis Hylton, Naval Facilities Engineering Command, Pacific, written commun., 2011). Here, it was assumed that all UFW for these systems was due to leakage.

Over half (24 Mgal/d) of the water supplied through the GWA system is unaccounted for (NAVFAC Pacific, 2010d). The amount of UFW that is due to leakage is not precisely known; estimates range from 10 percent to 40 percent of water input to the system (NAVFAC Pacific, 2010d). For this study, it was assumed that 25 percent of GWA's water production leaks from its water mains. Twenty-five percent was used in the Final Environmental Impact Statement (FEIS) analysis of projected future water demands and is the midpoint between GWA's estimate and the maximum estimate of the U.S. Environmental Protection Agency (NAVFAC Pacific, 2010d). Therefore, based on the 2010 average daily production rate (42 Mgal/d) (NAVFAC Pacific, 2010d), 10.5 Mgal/d of water is assumed to leak from the GWA water-main system. Because of the high uncertainty of actual leakage rates from the three water-main systems, recharge is also computed using different leakage rates (see "Sensitivity Analysis" section).

Water-main leakage was applied along the length of each water-main system. Leakage from Navy and Andersen AFB systems was applied at a uniform rate along the length of each water main. Leakage from the GWA system was applied according to water demand and water-main pipe volume. Of all the water supplied by the GWA system, about 80 percent is distributed to its northern and central systems (NAVFAC Pacific, 2010d). Therefore, it was assumed 80 percent of the water-main leakage occurs in the northern and central systems, and the remaining 20 percent of the leakage occurs in its southern system. Each GWA system is subdivided into villages. For each village, total pipe volume and the fraction of the system's total pipe volume was computed. Assuming that leakage is proportional to pipe volume, leakage for each village was applied at a rate that was proportional to its fraction of the total system pipe volume. For example, if a given village has 5 percent of the southern system's total pipe volume, this basin would incur 5 percent of the southern system's total water leakage; this equates to 1 percent (10.5 Mgal/d × 20 percent × 5 percent) or about 0.1 Mgal/d of the total leakage from the entire GWA water-main system.

Runoff

Direct runoff is the fraction of rainfall that does not contribute to net moisture gain within the plant-root zone (figs. 5 and 6). For this study, "direct runoff" is synonymous with "runoff" and is limited to water that flows over the land surface or within the shallow subsurface, and eventually discharges into Fena Valley Reservoir, surface streams that flow to the ocean, or directly to the ocean. Runoff does not include water that flows into closed depressions, including the Hagåtña Swamp area. Internally drained areas, which consist of closed depressions and their surface drainage basins, were determined in a GIS using a digital terrain model of the land surface (Joint Airborne Lidar Bathymetry

Technical Center of Expertise, unpublished) and following the procedure in Taylor and Nelson Jr. (2008).

Runoff-to-Rainfall Ratios

Runoff was estimated as a fraction of rainfall. For most of the study area, runoff was computed from daily stream-gaging records (southern Guam) or was assumed to be zero (northern Guam). Runoff from coastal urbanized areas in parts of Tumon, Tamuning, Hagåtña, and adjacent to Apra Harbor was accounted for separately (fig. 1) (see "Storm-Drain Systems" section). Streamflow measured by stream-gaging stations consists of direct runoff and base flow, which is groundwater that discharges into the stream. For each gaging-station record used in the analysis, direct runoff was separated from base flow using the hydrograph-separation program of Wahl and Wahl (1995) (see, for example, Gingerich, 2003). The direct-runoff component of streamflow was used in the computation of mean monthly runoff values for each stream gage.

Because the stream gages used in the runoff analysis measured streamflow during different periods, mean monthly runoff values for each gage were adjusted to a common period using "index gages." The period of the mean monthly rainfall maps, 1971–2000, was used as the common period. Stations 16847000, 16848500, and 16858000, on the Imong, Maulap, and Ylig Rivers, respectively, were used as index gages because of the completeness of their records. Linear equations relating monthly mean runoff values at a given index station to the other index stations were developed and were used to estimate missing monthly mean runoff values at the given index station. During a period in the mid-1990s when no continuous stream gages were operational on Guam, monthly mean runoff at the index stations was estimated using linear equations relating monthly mean runoff at each index station to monthly rainfall at weather station 914226.

For a given stream gage and index gage, mean monthly direct runoff was calculated for the gage in question and the index gage for the period when both gages were concurrently measuring streamflow. A coefficient of determination was computed between the mean monthly runoff values for the stream gage in question and the mean monthly runoff values for the index gage. This process was repeated for the other two index gages. An index gage was selected for the given stream gage according to (1) the coefficient of determination and (2) similarities in underlying geology. For the selected index gage, mean monthly runoff during the overlapping period was divided by mean monthly runoff for the period 1971–2000, resulting in 12 adjustment ratios. Each mean monthly runoff value for the non-index gage during the overlapping period was multiplied by the appropriate adjustment ratio to compute 12 adjusted mean monthly runoff values. The adjusted mean monthly runoff values for the gage were then divided by the mean monthly rainfall over its drainage basin (derived from Daly and Halbleib, 2006b). Thus, the computed monthly runoff-to-rainfall ratios are mean values for 1971–2000 (table 3). The daily runoff-to-rainfall ratio for a given month was assumed to be constant and equal to the monthly ratio.

Runoff Regions

Guam was divided into 11 runoff regions on the basis of geology, soils, topography, and known storm-drain systems (fig. 8). Some regions were subdivided according to drainage basins of stream-gaging stations. For each region with more than one stream-gaging station, each station's monthly runoff-to-rainfall ratios were applied over its entire drainage area; the remaining area of each region was assigned ratios that were averages of the other gages' ratios within the region. For regions with only one stream-gaging station, monthly runoff-to-rainfall ratios of the gage were applied over the entire region.

Runoff region 1 includes all areas where runoff to the ocean was assumed to be zero. The land surface of region 1 is mostly very permeable limestone and lacks well-developed stream channels that drain into the ocean. Except for steep cliffs, land-surface slope rarely exceeds 10 percent. Region 1 includes (1) most of the Orote Peninsula, (2) limestone areas and coastal lowlands along the southeast coast that are not within the drainage basin of any stream or river, (3) internally drained areas, and (4) most of the study area north of runoff regions 2 and 3.

Runoff regions 2 and 3 are mostly south of the Adelup fault (fig. 1) and primarily drain runoff from the Alutom Formation and parts of the southeast limestone plateau that are cut by stream channels. The land surface in these regions is steeply dissected and is covered by soils having slow to moderately slow permeability. Region 2 is east of the mountain ridgeline and includes drainage areas of most surface-water tributaries draining into Pago Bay and Ylig Bay. Region 3 is west of the mountain ridgeline and extends north from the Taleyfac River drainage basin to the Fonte River drainage basin. The western boundary of region 3 is defined where the slope of the land surface generally decreases to less than 10 percent.

Most of the Talofofo and Ugum River drainage basins, in central southern Guam, were divided into runoff regions 4, 5, and 6. Region 4 covers the northern portion of the interior basin and is defined as the drainage area of stream gage station 16845000. Region 4 is underlain by the Umatac and Alutom Formations and the Bonya and Alifan Limestones. In the part of this region characterized by cockpit karst morphology, streams sink into land surface and flow underground for tens to hundreds of meters before resurfacing. Region 5 is the drainage area of Fena Valley Reservoir. The land surface of region 5 is steeply sloped and is underlain by the Umatac Formation and older limestone. Region 6 includes the remaining areas of the Talofofo River drainage basin and most of the Ugum River drainage basin, but excludes the flat-lying valley floors. Predominantly underlain by Umatac and Alutom Formations, region 6 has steep ridges at higher elevations and gently sloped foothills at lower elevations.

Table 3. Ratios of runoff to rainfall used in the water-budget model for Guam.

[See figure 8 for locations of runoff regions; values of N, number of days, and f, turning point test factor, in the base-flow separation program (Wahl and Wahl, 1995) for all streams were 4 and 0.9, respectively; Avg., average; –, blank field]

Runoff region	Ratio of runoff to rainfall, adjusted to 1971–2000												
	Jan	Feb	Mar	Apr	May	June	July	Aug	Sept	Oct	Nov	Dec	Avg.
1	0.000	0.000	0.000	0.000	0.000	0.000	0.000	0.000	0.000	0.000	0.000	0.000	0.000
2a	0.261	0.383	0.238	0.123	0.404	0.159	0.336	0.541	0.449	0.496	0.477	0.337	0.350
2b	0.273	0.270	0.194	0.123	0.305	0.141	0.277	0.433	0.364	0.392	0.376	0.291	0.287
2c	0.267	0.327	0.216	0.123	0.354	0.150	0.307	0.487	0.406	0.444	0.427	0.314	0.319
3a	0.106	0.239	0.109	0.063	0.306	0.108	0.245	0.409	0.457	0.196	0.487	0.283	0.251
3b	0.147	0.171	0.182	0.073	0.375	0.094	0.214	0.280	0.272	0.310	0.259	0.198	0.215
3c	0.126	0.205	0.145	0.068	0.341	0.101	0.230	0.344	0.364	0.253	0.373	0.240	0.233
4	0.253	0.277	0.204	0.125	0.289	0.141	0.323	0.522	0.473	0.410	0.361	0.343	0.310
5a	0.164	0.220	0.101	0.094	0.222	0.134	0.227	0.389	0.350	0.353	0.330	0.263	0.237
5b	0.171	0.248	0.100	0.095	0.210	0.134	0.194	0.372	0.335	0.302	0.388	0.266	0.235
5c	0.168	0.234	0.100	0.094	0.216	0.134	0.210	0.380	0.343	0.328	0.359	0.264	0.236
6	0.147	0.242	0.074	0.060	0.175	0.092	0.147	0.221	0.186	0.220	0.213	0.186	0.164
7	0.132	0.193	0.069	0.067	0.221	0.071	0.133	0.272	0.231	0.248	0.224	0.160	0.168
8	0.189	0.227	0.093	0.107	0.295	0.095	0.194	0.362	0.298	0.306	0.287	0.209	0.222
9a	0.296	0.394	0.165	0.181	0.402	0.224	0.312	0.411	0.325	0.355	0.378	0.386	0.319
9b	0.201	0.329	0.186	0.129	0.577	0.110	0.248	0.415	0.377	0.340	0.467	0.267	0.304
9c	0.179	0.233	0.122	0.106	0.284	0.082	0.232	0.339	0.293	0.328	0.310	0.286	0.233
9d	0.226	0.319	0.158	0.139	0.421	0.139	0.264	0.389	0.332	0.341	0.385	0.313	0.285
10	0.132	0.193	0.069	0.067	0.221	0.071	0.133	0.272	0.231	0.248	0.224	0.160	0.168
11	0.000	0.000	0.000	0.000	0.000	0.000	0.000	0.000	0.000	0.000	0.000	0.000	0.000

Region 7, near the eastern coastline of southern Guam, includes the drainage basins of the Togcha, Asalonso, and Pauliluc Rivers. The Togcha and Asalonso Rivers cut through the limestone plateau, and the Pauliluc River drains gently-sloping foothills of the Umatac Formation. The land surface in region 7 has a wide range of slopes and is underlain by soils having slow to moderately rapid permeability. Region 8 includes most of the drainage basins of the surface-water networks that flow into Inarajan Bay and Agfayan Bay. Underlain by the Umatac Formation, the land surface of region 8 is steep at higher elevations and near stream channels, but is relatively flat in the north. Region 9 consists of the drainage basins of all surface-water channels between the Ajayan River and Sagua River, but excludes coastal lowlands. The land surface of region 9 includes Facpi and Umatac Formations and is very steep.

Region 10 includes coastal lowlands and valley floors in southern Guam where the land surface has slopes of less than 10 percent. Derived from alluvium and volcanics, the soils in region 10 have slow to moderately rapid permeability. Runoff-to-rainfall ratios from region 7 were applied to region 10 because the land surface slope and soils of region 10 are more similar to region 7 than any other region.

Region 11 includes coastal urbanized areas where runoff is collected by storm-drain systems and is routed to the ocean. The pervious parts of region 11 are mainly underlain by limestone and alluvium, where runoff is expected to be low. Therefore, runoff region 11 was assigned monthly runoff-to-rainfall values of zero. Runoff from the impervious parts of this region, however, is accounted for separately (see Storm-Drain Systems).

Storm-Drain Systems

Some urbanized areas use storm-drain systems to dispose of rain that falls onto impervious surfaces. A typical storm-drain system consists of inlets, conduits, manholes, and appurtenances that collect and carry water from impervious surfaces to a point of discharge (U.S. Army Corps of Engineers, 1980). Rainfall collected by storm-drain systems is commonly referred to as "stormwater runoff."

Table 3. Ratios of runoff to rainfall used in the water-budget model for Guam—Continued

[See figure 8 for locations of runoff regions; values of N, number of days, and f, turning point test factor, in the base-flow separation program (Wahl and Wahl, 1995) for all streams were 4 and 0.9, respectively; Avg., average; –, blank field]

Runoff region	Method used to calculate runoff-to-rainfall ratios	Gaging-station number	Stream/river/creek	Basin area (acres)	Periods of record used in calculation
1	Assume runoff is zero	–	–	–	–
2a	USGS stream-gaging station	16865000	Pago	3,565	1951–83, 2000–10
2b	USGS stream-gaging station	16858000	Ylig	4,177	1952–95, 1997–2001
2c	Average of runoff regions 2a and 2b	–	–	–	–
3a	USGS stream-gaging station	16807650	Aplacho	279	1999–2004, 2006–10
3b	USGS stream-gaging station	16808300	Finile	169	1960–82
3c	Average of runoff regions 3a and 3b	–	–	–	–
4	USGS stream-gaging station	16845000	Tolaeyuus	4,044	1951–1960
5a	USGS stream-gaging station	16848500	Maulap	755	1972–94, 1997–2010
5b	USGS stream-gaging station	16847000	Imong	1,228	1960–94, 1997–2010
5c	Average of runoff regions 5a and 5b	–	–	–	–
6	USGS stream-gaging station	16855000	Ugum	4,538	1952–1970
7	USGS stream-gaging station	16840000	Tinaga	1,220	1952–1985
8	USGS stream-gaging station	16835000	Inarajan	2,780	1952–1983
9a	USGS stream-gaging station	16809400	Cetti	473	1960–1967
9b	USGS stream-gaging station	16809600	La Sa Fua	659	1953–60, 1976–84, 2000–10
9c	USGS stream-gaging station	16816000	Umatac	1,332	1952–76, 2002–09
9d	Average of runoff regions 9a, 9b, and 9c	–	–	–	–
10	Assume same as runoff region 7	–	–	–	–
11	Runoff accounted for separately	–	–	–	–

This study accounts for stormwater runoff that is collected by storm-drain systems and is discharged to the ocean, the Harmon Sink, or drywells. West of the Guam International Airport is the Harmon Sink, a large surface depression more than 10 acres in area, that is underlain by limestone (fig. 9). The Harmon Sink receives stormwater runoff from much of the surrounding urbanized areas, including street drainage from a storm-drain network (NAVFAC Pacific, 2010b) and parts of the Guam International Airport (Moran and Jenson, 2004). Drywells are boreholes in the ground, typically one to two feet in diameter, and tens to hundreds of feet deep (Earth Tech, Inc., 1999). Drywells collect overland runoff or receive channeled (drainage pipe or ditch) flow.

The study area was divided into four storm-drain zones according to where stormwater runoff is routed (fig. 9). Storm-drain zone 1 includes coastal urbanized areas inland from Hagåtña Bay and Tumon Bay. Stormwater runoff from these areas is routed to the ocean by way of storm drains or the Tamuning drainageway. The spatial extent of this zone was estimated using maps and descriptions in the "Guam

Storm Drainage Manual" (U.S. Army Corps of Engineers, 1980) and the Stormwater Implementation Plan for the Guam Road Network (NAVFAC Pacific, 2010b). Areas adjacent to Apra Harbor that are mostly impervious were also included in storm-drain zone 1 because stormwater runoff from these areas likely flows directly into the ocean.

Storm-drain zone 2 includes parts of Andersen Air Force Base (AFB) and the Guam International Airport where much of the stormwater runoff is routed into drywells (Ogden Environmental and Energy Services, Inc., 1995; Earth Tech, Inc., 1999; Brian Ho, AECOM, written commun., 2011). The cumulative drainage area of all drywells at Andersen AFB, documented in Earth Tech, Inc. (1999), was used to define the part of zone 2 covering Andersen AFB. At the Guam International Airport, storm-drain zone 2 encompasses the runways, taxiways, and impervious areas east of the runways.

Storm-drain zone 3 includes the cumulative drainage area of the storm-drain systems that route stormwater runoff into the Harmon Sink. The spatial extent of zone 3 was estimated using descriptions in U.S. Army Corps of Engineers (1980),

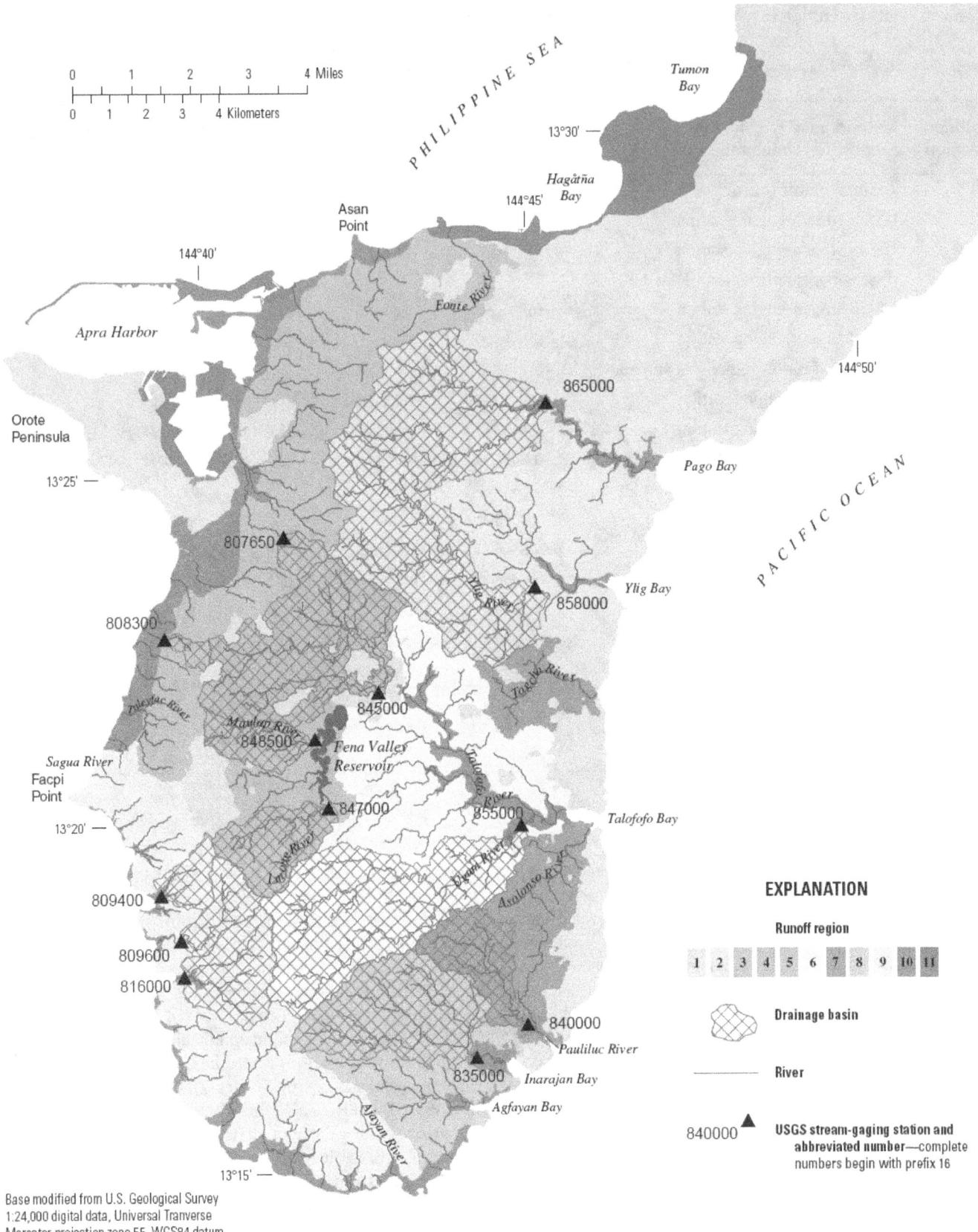

Figure 8. Runoff regions defined for the water-budget calculation on Guam. All areas where runoff was assumed to be zero (runoff region 1) are shown in light gray. The part of the study area not shown on this map lies within runoff region 1.

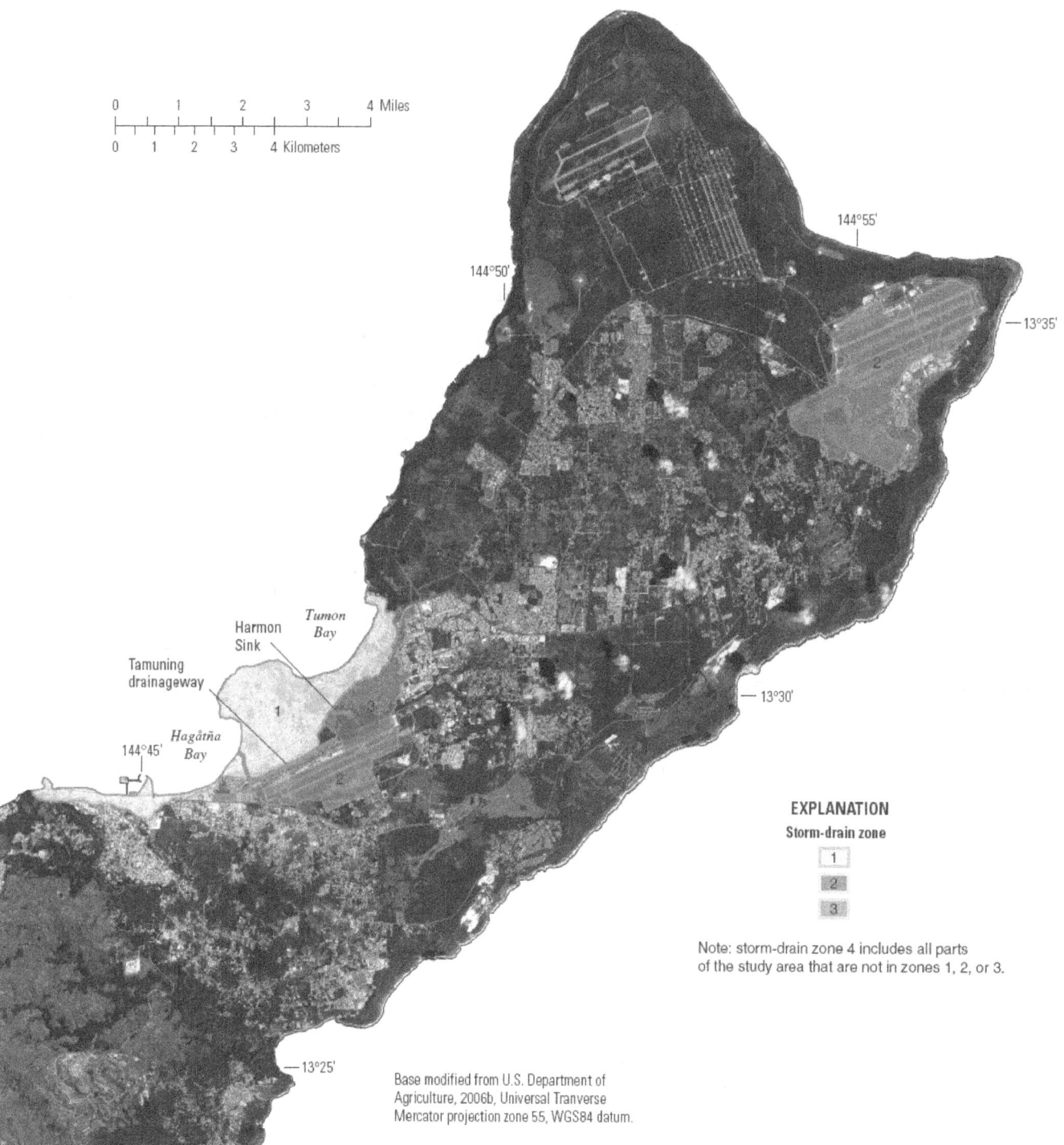

Figure 9. Storm-drain zones defined for the water-budget calculation on Guam.

Moran and Jenson (2004), NAVFAC Pacific (2010b), and a digital elevation model of the land surface.

Storm-drain zone 4 includes all other areas in the study area that are not in storm-drain zones 1, 2, or 3. Although urbanized areas within zone 4 are known to have storm-drain systems, these areas were not defined as unique storm-drain zones because information describing their catchment areas and disposal locations is not available. For the water-budget model, it was assumed that 0.25 in. of incident rainfall can be stored on an impervious surface in surface depressions and potholes. This is the rainfall interception capacity, N, of impervious surfaces, which is introduced in equations 2 and 3. An N value of 0.25 in. was used in previous water budgets for Hawaiʻi (Izuka and others, 2005; Engott and Vana, 2007; Engott, 2011). Rainfall in excess of the interception capacity is assumed to be stormwater runoff. For storm-drain zone 4, all stormwater runoff was added to the pervious fraction of the subarea. For storm-drain zones 1, 2, and 3, a fraction of the stormwater runoff within each zone was assumed to be captured by the storm-drain system, and the uncaptured stormwater runoff was added to the pervious fraction of the subarea. Heitz and others (1997) recommended that ponding basins on Guam should be designed to capture between 60 and 80 percent of the runoff within their respective catchment areas. Here, for each subarea within storm-drain zones 1, 2 and 3, it was assumed that 50 percent of the stormwater runoff was captured by the storm-drain systems; the remaining stormwater runoff that was not captured by the storm-drain systems was added to the pervious fraction of each subarea. The captured stormwater runoff for storm-drain zone 1 was counted as runoff to the ocean. Because drywell boreholes are typically cased to depths several feet below the land surface, which is generally below the rooting depth of most plants, captured runoff in storm-drain zone 2 was counted as direct recharge, DR (see equation 13). The captured runoff was applied evenly to all areas within zone 2, except for areas classified as urban builtup. Stormwater runoff captured in storm-drain zone 3, which is routed to the Harmon Sink, also was counted as direct recharge. In storm-drain zones 1, 2, and 3, evapotranspiration of captured stormwater runoff was assumed to be zero.

Evapotranspiration

Evapotranspiration (ET) is the sum of all water that is evaporated or transpired from the plant-soil system. Evapotranspiration can be subdivided into three main evaporative processes: (1) canopy evaporation, which is evaporation of intercepted rain from the surface of vegetation; (2) ground evaporation, which is evaporation from the soil surface and overlying litter and mulch layers; and (3) transpiration, the process by which soil moisture taken up by vegetation is eventually evaporated as it exits at plant pores (Viessman and Lewis, 2003, p. 143). Recognizing that these evaporative processes tend to operate on different time scales ranging from hours (canopy evaporation and ground evaporation) to weeks (transpiration) (Savenije, 2004), Engott (2011) developed a method to separately quantify these evaporative processes for certain land covers in a daily water-budget model for the island of Hawaiʻi. In contrast, none of these evaporative processes were quantified individually in any of the previous water-budget investigations on Guam (table 1). In this study, following the method developed by Engott (2011), ET in forests is calculated by separately estimating canopy evaporation and combined ground evaporation and transpiration. These two terms are then added together to yield a total ET rate. For all other land covers, ET is calculated in which no separate estimates of canopy evaporation and combined ground evaporation and transpiration are made. The concept of potential ET, combined with soil-moisture limiting, is used to estimate ground evaporation and transpiration in forests and total ET for all other land covers.

Canopy Evaporation and Net Precipitation

As rain falls on a vegetated surface, a fraction of the droplets will strike and collect on the leaves, trunks, or stems of the vegetation in a process known as canopy interception. The fraction of rainfall that remains on the vegetation, and ultimately evaporates, is called "canopy evaporation." The remaining fraction of rainfall that ultimately reaches the soil surface is called "net precipitation" (fig. 5) (Crockford and Richardson, 2000).

Canopy evaporation in forested areas is an important parameter to include in any water-budget study (McJannet and others, 2007). Numerous studies indicate that canopy evaporation in tropical forests may substantially reduce the rainfall that reaches the ground beneath a forest canopy compared to areas that are not forested (Hutjes and others, 1990; Dykes, 1997; Asdak and others, 1998). Because of the height of trees, turbulent diffusion is much more efficient at removing intercepted water from forests than from other land-cover types; this enhanced rate of evaporation from a wet forest canopy makes realistic estimates of ET from forests possible only if transpiration and canopy evaporation are evaluated separately (Shuttleworth, 1993).

Because canopy evaporation is difficult to directly measure, it is typically estimated from net precipitation measurements. Net precipitation is determined by measuring the precipitation that reaches the floor of a forest, beneath the canopy, and comparing it to rainfall collected contemporaneously above the forest canopy or in a nearby open field. Therefore, net precipitation is typically reported as a percentage of rainfall, and canopy evaporation is estimated as the difference between gross rainfall and net precipitation.

The amount of precipitation that reaches the ground beneath a forest canopy depends on forest type and climatic factors (Crockford and Richardson, 2000). Because no known net precipitation studies have been conducted on Guam, net precipitation in forested areas for this study was estimated from published measurements in other regions that have similar forests and climate. Bidin and Chappell (2004) measured net precipitation beneath a disturbed, secondary forest in northeast Borneo that had been selectively logged 8 years prior to the study. Net precipitation beneath forest canopies that had been regenerated by various types of vines, sprawlers, and other pioneer species ranged from 82 to 88 percent of annual rainfall, and averaged 85 percent. Asdak and others (1998) measured net precipitation beneath tropical forests in Indonesia. Rainfall at the study site is seasonally variable, with most rainfall being convective in origin and occurring in heavy downpours. At the time of the measurements, the stature of the trees comprising the logged portion of the forest was similar to trees on Guam. Net precipitation beneath this portion of the forest that had a closed canopy was 85 percent of annual rainfall. On the basis of the results of Asdak and other (1998) and Biden and Chappell (2004), net precipitation beneath all forests on Guam was assumed to be 85 percent of annual rainfall.

Owing to seasonal differences in rainfall intensity, monthly and daily net-precipitation values beneath forests can be much different than annual net-precipitation values. For example, at a lowland forest in northeast coastal Australia where annual net precipitation was 26 percent of annual rainfall, monthly net precipitation ranged from less than 20 to more than 80 percent of monthly rainfall (McJannet and others, 2007). At this site, months with relatively high-intensity rainfall had greater net precipitation, whereas months with relatively low-intensity rainfall had less net precipitation. On a daily basis, measurements by McJannet and others (2007) and Bruijnzeel and Wiersum (1987) indicate that net precipitation was linearly related to rainfall intensity. On the basis of these observations and because the Guam water-budget operates on a daily time step, daily net precipitation, in inches, was calculated using the following equation (derived from Wallace and McJannet, 2006):

$$(NP)_i = m \times P_i - CS, \tag{16}$$

where

$$
\begin{aligned}
(NP)_i &= \quad \text{net precipitation for the day [inches],}\\
m &= \quad \text{daily constant [dimensionless],}\\
P_i &= \quad \text{precipitation for the day [inches],}\\
CS &= \quad \text{canopy storage capacity [inches].}
\end{aligned}
$$

In equation 16, the canopy storage capacity (CS) is assumed to be the maximum possible amount of water that can be stored in the forest canopy. For days when the product of the daily constant and rainfall, $m \times P_i$, is less than or equal to the canopy storage capacity, net precipitation is assumed to be zero. Canopy storage capacity is assumed to be 0.89 mm (0.035 in.), which is the average of the canopy-storage capacity values reported by Jackson (1975), Calder and others (1986), Bruijnzeel and Wiersum (1987), Lloyd and others (1988), and Schellekens and others (1999). To determine the

value for the daily constant (m) that is needed to produce a net annual precipitation value that is 85 percent of annual precipitation in forested areas, the recharge model was run multiple times with m incrementally changed for each successive run. The values of all other input parameters were set to baseline conditions, which used 2004 land cover and mean annual rainfall for 1961–2005, and were not altered for successive model runs. The m value that produced an annual net precipitation value of 85 percent in forested areas was 1.095.

Potential Evapotranspiration

Potential evapotranspiration is the maximum rate that water can be removed from the plant-root zone by ET if soil moisture is nonlimiting (Giambelluca, 1983). The actual rate of ET becomes less than the potential rate with the onset of soil-moisture stress. As the soil dries, capillary and adsorptive forces bind the remaining water to the soil matrix more strongly, reducing water flow to roots. Soil-moisture stress occurs when the decreasing flow of water to the root system induces a response in the plant to slow down transpiration and prevent desiccation. The threshold-moisture content at which a plant begins to react to soil drying varies with the type of plant. The actual rate of ET is a function of potential ET, soil-moisture content, and threshold-moisture content (see equation 12).

Potential evapotranspiration is controlled by prevailing atmospheric conditions (solar radiation, air temperature, humidity, and wind) and land-cover characteristics (reflectance, roughness, and plant physiology) (Giambelluca, 1983). Potential-ET measurements on Guam are almost nonexistent. Previous recharge studies on Guam (Mink, 1976; Jocson and others, 2002; Habana and others, 2009) derived potential ET from pan evaporation measurements at weather station 914229, in the northern half of the island (fig. 2). Pan evaporation data from weather station 914229 was not used in this study because pan-evaporation measurements at this station were discontinued in 1996.

Instead of using pan evaporation, reference evapotranspiration was used as the basis for determining potential ET. Also referred to as reference-crop evapotranspiration, reference ET is the evapotranspiration rate from a hypothetical grass surface of specific characteristics, with no soil-moisture stress (Allen and others, 1998). Calculated from weather measurements, reference ET indicates the evaporative power of the atmosphere at a given place and time, but is independent of land-cover characteristics (Allen and others, 1998). To account for land-cover characteristics, crop coefficients were developed on the basis of vegetation cover. A crop coefficient is an empirically derived ratio of potential ET to reference ET. Thus, for each land-cover class (fig. 4), potential ET is estimated by multiplying reference ET for the area by the appropriate crop coefficient. In contrast, none of the previous water budgets (table 1) accounted for the potential-ET rate of each land-cover type in their respective study areas.

Reference Evapotranspiration

Reference ET was computed using the Food and Agricultural Organization (FAO) Penman-Monteith method in Allen and others (1998). The FAO Penman-Monteith equation requires air temperature, air humidity, wind speed, and solar radiation data. Daily maximum air temperature, minimum air temperature, dew-point temperature, and wind-speed data were obtained from the National Climate Data Center. Daily solar-radiation values were computed from hourly solar-radiation data obtained from the National Solar Radiation Database (National Renewable Energy Laboratory, 2007).

Reference ET was calculated for weather station 914226 using daily weather observations from 1961 to 2005 (fig. 7). This time period was limited to the availability of solar-radiation data. Weather station 914226 only has solar-radiation data from 1991 to 2005. Therefore, following the recommendation of Allen and others (1998), solar-radiation data from weather station 914229 during 1961–1990 were used. There were no missing solar-radiation values in the two datasets. Of the daily air temperature, dew-point temperature, and wind-speed observations used in the reference ET calculation, more than 98 percent were from weather station 914226, and the remaining weather observations were estimated from observations at weather stations 914025 and 914229. During 1961–2005, only two years of record from station 914226 were missing more than 10 percent of the daily air temperature, dew-point temperature, and wind-speed observations (fig. 7). To help estimate missing weather observations at station 914226, linear-regression equations relating daily weather observations at this station to contemporaneous weather observations at stations 914025 and 914229 were developed, and Spearman's correlation coefficients for the concurrent observations were calculated. Missing air-temperature and dew-point temperature values at weather station 914226 were estimated using linear-regression equations and data from weather station 914229 because weather observations from station 914229 were more highly correlated to observations from station 914226 than were weather observations from station 914025. Linear-regression equations and data from weather station 914025 were used to estimate air-temperature and dew-point temperature values for days when stations 914229 and 914226 had missing values. Following the recommendation of Allen and others (1998), wind-speed values from weather stations 914229 and 914025 were used in the calculation of reference ET for days when weather station 914226 did not have wind observations. Wind-speed values from weather station 914229 were preferentially used over values from weather station 914025 because wind-speed values from station 914229 were more highly correlated to values from station 914226 than were wind-speed values from station 914025.

The calculated reference ET for station 914226 was applied to the entire study area because (1) the daily weather-observation values required for the FAO Penman-Monteith equation at stations 914025 and 914229 were, on average, within 10 percent of contemporaneous weather-observation values at station 914226, and (2) there is an absence of other weather stations on the island with long-term, reliable weather observations. For 1961–2005, calculated daily reference ET ranged between 0.03 and 0.31 in. Mean monthly reference ET ranged from 0.16 inches per day (in/d) in October to 0.20 in/d in April. Annual reference ET ranged between 57.9 and 69.4 in., and averaged 63.2 in. (fig. 7). Variations in annual reference ET during this time period were accounted for in the water-budget model using the same method applied to rainfall. Annual reference ET during the 45-year time period was divided by the mean annual reference ET, creating a series of 45 weighting factors. Reference ET for a given month was calculated by multiplying the mean monthly value by the weighting factor appropriate for the year. Daily reference ET for a given month was assumed to be constant and was equal to monthly reference ET divided by the number of days in the month.

Crop Coefficients

Crop coefficients are ratios of potential ET to reference ET for a given land cover. No known studies have been conducted on Guam to quantify potential-ET rates of vegetation on the island. Previous water-budget studies on Guam did not estimate the potential-ET rate of each land-cover type in their respective study areas. In contrast, this study assigns rates of potential ET to each land-cover type using crop coefficients. Crop coefficients were derived from published ET studies in other areas with similar vegetation.

Crop coefficients for forest land covers were estimated as the combination of ground evaporation and transpiration. Forest crop coefficients incorporate ET estimates of two of the most abundant tree species on Guam—coconut palms and *L. leucocephala*—which together represent about 25 percent of the gross volume of the island's forests (Donnegan and others, 2004). Forest crop coefficients are also derived from reported ET rates for tropical forests that are representative of multiple tree species (not species-specific). Unless estimates were provided in these studies, daily ground evaporation was assumed to be 0.1 mm (0.004 in.), based on measurements by Jordan and Heuveldop (1981) for a tropical rainforest in Venezuela.

Roupsard and others (2006) monitored ET of coconut palms for 3 years on the south Pacific island of Vanuatu. Using sapflow measurements, average annual palm transpiration was estimated to be 642 mm (25.3 inches), which was about 45 percent of average annual reference evapotranspiration. Including an assumed annual ground evaporation of 1.5 inches (0.004 in/d × 365 d), results in a crop coefficient of 0.48. Because Roupsard and others (2006) reported that the soil profile remained close to field capacity during the study, it is assumed transpiration rates were not reduced by soil-moisture stress.

Das and others (1990) estimated transpiration of *L. leucocephala* stands in India using soil-moisture measurements. During a 41-day wet period (December to January), the total transpiration of *L. leucocephala* was 49.96 mm (1.97 inches). The total pan evaporation during this period was 63.5 mm (2.50 inches). Measurements from this period were used because rainfall was greater than pan evaporation, and soil-moisture stress was assumed to be negligible. Measurements from the remainder of the study period were not used because pan evaporation exceeded rainfall during most of this period, and the resulting soil-moisture deficit likely reduced transpiration. Adding an assumed 0.16 inches of ground evaporation (0.004 in/d × 41 d) to 1.97 inches equates to 2.13 inches of combined ground evaporation and transpiration; this is 85 percent of pan evaporation (2.13 / 2.50 inches). On Guam, concurrent measurements of pan evaporation at weather station 914229 and reference ET at weather station 914226 indicate that reference ET is about 82 percent of pan evaporation. Therefore, 0.85 divided by 0.82 computes to a crop coefficient of 1.04 for *L. leucocephala*.

Calder and others (1986) monitored soil moisture to estimate transpiration of a disturbed, secondary lowland tropical forest in West Java, Indonesia. Annual transpiration was estimated to be 886 mm (34.9 inches). Calder and others (1986) reported that the soil-moisture deficit during the study period, even when at its maximum, was not large enough to substantially reduce transpiration rates. Annual net radiation was 3.73×10^9 Joules, which is equivalent to 1,520 mm (59.8 inches) of water evaporation using the conversion 10^6 Joules/m^2/d = 0.408 mm/d from Allen and others (1998, table 1). Adding an assumed 1.5 inches of annual ground evaporation to 34.9 inches equates to 36.4 inches of combined ground evaporation and transpiration; this is 61 percent of annual net radiation. At weather station 914226 on Guam, mean annual reference ET, calculated using equation 6 in Allen and others (1998), is 95 percent of mean annual net radiation, calculated using equation 40 in Allen and others (1998) (see "Reference Evapotranspiration" section). Therefore, 0.61 divided by 0.95 computes to a crop coefficient of 0.64 for the forest.

McJannet and others (2007) measured sapflow to estimate transpiration of tropical rainforest sites in northeast Australia. At the coastal site (Oliver Creek) where water input from fog was absent, monthly crop coefficients for all months except September to November averaged 0.60. This crop coefficient is calculated as the sum of transpiration and ground evaporation, divided by reference ET, as depicted in figure 4 of McJannet and others (2007). Crop coefficient values for the months of September to November were not used because low soil moisture during these months may have resulted in reduced forest transpiration at this site.

Forest crop coefficients used in the Guam water budget (table 2) are computed averages of the forest crop coefficients derived from (1) Roupsard and others (2006), (2) Das and others (1990), (3) Calder and others (1986), and (4) McJannet and others (2007). The crop coefficient assigned to limestone forest and scrub forest is the average of all four studies. The crop coefficient assigned to ravine forest is the average of all studies except Das and others (1990) because *L. leucocephala* grows well on limestone soils, but is scarce on volcanic soils (Fosberg, 1960). The crop coefficient from Das and others (1990) was assigned to *Leucaena* stand. Coconut plantation was assigned the crop coefficient derived from Roupsard and others (2006). A crop coefficient of 0.64 was assigned to both *Acacia* plantation and *Casuarina* thicket and is the average of Calder and others (1986) and McJannet and others (2007).

For nonforest land covers, crop coefficients were based on total evapotranspiration. Crop coefficients for urban cultivated, barren land, golf course, agricultural field, wetland, and water were derived from mid-season crop coefficients listed in Allen and others (1998). The crop coefficient listed for warm-season turfgrass was used for urban cultivated and golf course. The crop coefficient listed for bare soil was used for barren land. The average of the crop coefficients for common agricultural crops on Guam was used for agriculture field. Common agricultural crops were determined from Young (1988). Crop coefficients derived from Allen and others (1998) were adjusted to climatic conditions on Guam using equation 62 in Allen and others (1998) and using an average daily wind-speed value of 6.93 miles per hour (adjusted to a measurement height of 6.56 ft) and an average minimum relative humidity of 71 percent.

The crop coefficient for urban builtup was based on the pan coefficient, 1.0, for the urban category in Giambelluca (1983). Similar to a crop coefficient, a pan coefficient is the ratio of potential ET to pan evaporation for a given land cover. On Guam, reference ET is about 82 percent of pan evaporation. Dividing the pan coefficient, 1.0, by 0.82 results in a crop coefficient of 1.22.

The crop coefficient for savanna complex was derived from evapotranspiration estimates of *Miscanthus giganteus*. On Guam, savanna complex is dominated by a similar plant, *Miscanthus floridulus*, also known as sword grass (Fosberg, 1960). On the basis of energy-balance measurements in Illinois, Hickman and others (2010) determined the latent heat flux (energy consumed for ET) of *M. giganteus* was about 14 percent greater than maize (corn) when both plants had a mature canopy. During this period, air temperature, precipitation, and soil-moisture conditions at the study site were "normal" as assessed by the Palmer Crop Moisture Index (Hickman and others, 2010). Suyker and Verma (2009) estimated a mid-season crop coefficient of 1.03 for irrigated maize in Nebraska using eddy covariance flux measurements. Allen and others (1998) list a mid-season crop coefficient of 1.20 for maize (field corn). Multiplying 1.12, the average of 1.03 and 1.20 for maize, by 1.14 results in a crop coefficient of 1.28 for *M. giganteus*. The studies of Suyker and Verma (2009) and Hickman and others (2010) were both conducted in the midwest region of the United States, and it was assumed the climatic conditions were similar to standard conditions in Allen and others (1998). Accordingly, the derived crop coefficient for *M. giganteus* was adjusted to 1.23 using equation 62 in Allen and others (1998).

Moisture-Storage Capacity

Moisture-storage capacity (fig. 10) was computed as the product of available water capacity and root depth (equation 10). Available water capacity varies by soil type and is a measure of the maximum depth of water per unit depth of soil available for consumption by plants. The U.S. Department of Agriculture (2009) soil map and corresponding tables of available water capacities were used to distribute available water capacity over the study area. The tables list available water capacities for each soil type as minimum and maximum values at various ranges of depth. For this study, a depth-weighted average available water capacity was computed and assigned to each soil type.

Each land-cover category was assigned a root depth that was estimated from published information for similar types of vegetation and land cover (table 2). Using a compilation of published root studies, Jackson and others (1996) estimated the parameters for the vertical root distribution model of Gale and Grigal (1987), which is used to estimate the cumulative proportion of roots at different soil depths for major terrestrial biomes. For this study, root depths derived from the results of Jackson and others (1996) represent the soil depth estimated to contain about 85 percent of a particular biome's roots. Root depths for all forest land covers are based on tropical evergreen forest in Jackson and others (1996). Root depth for savanna complex is the average of the tropical savanna value derived from Jackson and others (1996) and the value reported for *Miscanthus* in Beale and others (1999). Root depths for urban cultivated and urban builtup are based on the urban category in Giambelluca (1983). Root depth for other shrub/grass is derived from the sclerophyllous shrub biome in Jackson and others (1996). Root depth for golf course is based on the warm-season turfgrass listing in Allen and others (1998). The root depth for barren land is the average of the high and low values suggested by Allen and others (1998) for bare soil. The root depth for agriculture field is based on diversified agriculture in Fares (2008). The root depth for wetland is the same as wetland vegetation in Engott (2011).

Other Input

In addition to the water-budget inputs already discussed, several other inputs are required. The initial moisture storage was set at 50 percent of capacity, and the annual rate of groundwater recharge from surface-water bodies was set at 12 in. The values assigned to these parameters are consistent with those for recent Hawai'i water budgets (Izuka and others, 2005; Engott and Vana, 2007; Engott, 2011). The effects of these parameters on regional-scale recharge generally are minor because they either pertain to only a small area or are applicable during only a small fraction of time.

Model Randomness

The selection of monthly rainfall fragment sets (see "Rainfall" section) introduces randomness into the water-budget model. To account for this randomness, the water-budget model was run multiple times, and the results were averaged. To determine the appropriate number of simulations to run, the water-budget model for Guam was run 20 times.

The marginal, absolute percentage change in groundwater recharge for each of the 138,408 subareas was averaged for each number of simulations (fig. 11). After 10 simulations, the average percentage change did not exceed 0.05 percent. This very small value, 0.05 percent, was determined to be adequate for this study. Accordingly, for the baseline and future land-cover scenarios, the water-budget model was run 10 times, and the results were averaged.

Groundwater-Recharge Estimates Using the Water-Budget Model

Baseline Recharge

For the entire island of Guam, estimated mean annual recharge is 394.1 Mgal/d for baseline conditions (table 4). Baseline conditions for this study were 2004 land cover and mean annual rainfall and evapotranspiration during 1961–2005. Recharge is 39 percent of mean annual rainfall (999.0 Mgal/d). Compared to rainfall for the entire study area, water inflows from irrigation (0.8 Mgal/d), septic systems (4.8 Mgal/d), and water-main leakage (12.8 Mgal/d) are relatively minor. Of the total water inflow, 13 percent becomes runoff, and 49 percent becomes ET and canopy evaporation.

Although minor in comparison to rainfall for the entire island, water-main leakage, septic-system leachate, and stormwater runoff can add substantial amounts of water to certain subareas. At the approximately 15,000 localities in the study area defined as septic systems, mean annual water input to the soil from septic systems is between 2 and 3 times greater than mean annual rainfall. For subareas with water-main pipes, annual water-main leakage is between 0.4 and 4 times greater than mean annual rainfall. For subareas at Andersen AFB and Guam International Airport that receive stormwater runoff, mean annual water input from stormwater runoff is between 25 and 30 percent of mean annual rainfall. At the Harmon Sink, where storm-drain systems dispose of stormwater runoff from the surrounding urbanized area, annual water input from stormwater runoff is 0.7 Mgal/d, which is about 7 times greater than mean annual rainfall.

For subareas that only receive rainfall and do not receive supplemental water from water-main leakage, septic-system leachate, or stormwater runoff, maximum annual recharge is 78.7 inches. Therefore, all subareas where annual recharge is greater than 78.7 inches because of water inflow from one or more of these supplemental water-inflow components are classified here as having "artificially high recharge." Although recharge is less than 78.7 inches per year at some subareas that receive water-main leakage or stormwater runoff, 78.7 inches per year was set as the minimum value for artificially high recharge in order to simplify displaying recharge distribution (fig. 12). The maximum value for artificially high recharge is 768 inches per year at the Harmon Sink.

In general, recharge is highest in areas underlain by limestone and is lowest in areas underlain by volcanic soils

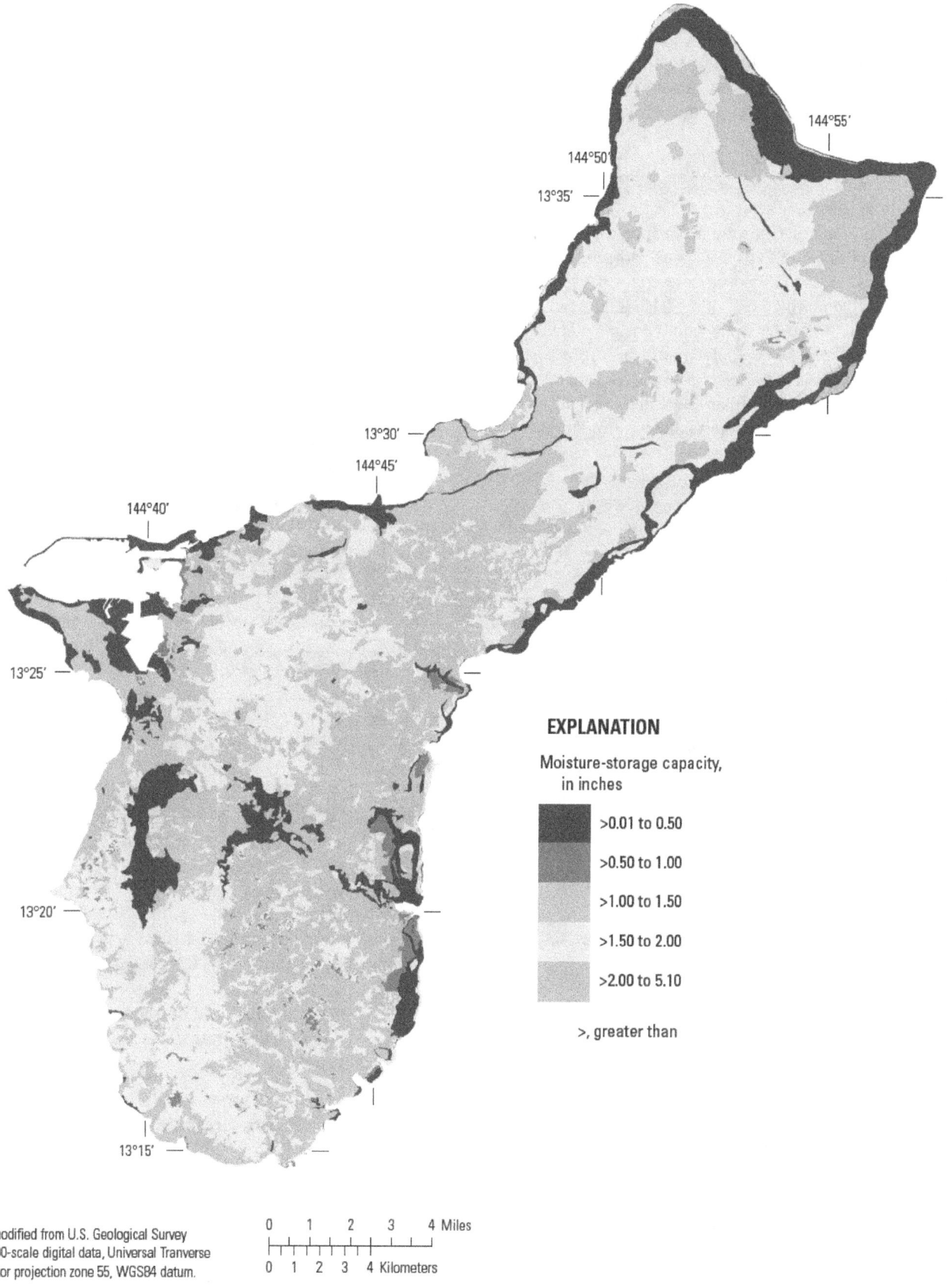

Figure 10. Calculated moisture-storage capacity on Guam.

EXPLANATION

Moisture-storage capacity,
in inches

>0.01 to 0.50

>0.50 to 1.00

>1.00 to 1.50

>1.50 to 2.00

>2.00 to 5.10

>, greater than

Base modified from U.S. Geological Survey
1:24,000-scale digital data, Universal Tranverse
Mercator projection zone 55, WGS84 datum.

0   1   2   3   4 Miles

0   1   2   3   4 Kilometers

(fig. 12). Most recharge occurs in the northern half of the island, which is predominantly underlain by thin, permeable limestone soils, and has almost no runoff. The total mean annual recharge for the northern aquifer sectors defined by Mink (1991), which encompass most of the northern half of Guam, is 238.0 Mgal/d, or about 51 percent of mean annual rainfall (table 4). Here, of the total water inflow, only 1 percent becomes runoff, and 49 percent becomes ET and canopy evaporation. Mean annual recharge for the northern aquifer sectors, expressed as a uniform depth spread over each sector's area, ranges from 47.5 to 51.2 in. per sector, and is 48.9 in. for the combined areas of all northern sectors.

Across the northern half of the island, the spatial variability is fairly minor. Local recharge maxima include the parts of Andersen AFB and the Guam International Airport area where stormwater runoff is routed into drywells. In general, recharge is greater in grassy areas (urban cultivated) than in forested areas (scrub and limestone forests) having similar rainfall owing to canopy evaporation in forested areas. Compared to other areas in the northern half of the island, areas covered by continuous stands of *L. leucocephala* (Leucaena stand) have the lowest recharge, owing to canopy evaporation and relatively high ET. Recharge is also relatively low in the Tumon and Tamuning areas where stormwater runoff is captured by storm-drain systems and routed to the ocean, and in the urbanized areas inland from Tumon Bay where stormwater runoff is routed to the Harmon Sink.

The part of the study area that encompasses most of the southern half of the island and is not within the aquifer sectors defined by Mink (1991) is defined as the "southern" aquifer sector for this report. For the southern aquifer sector, mean annual recharge is 156.1 Mgal/d, which is 30 percent of mean annual rainfall (table 4). Here, of the total water inflow, 23 percent becomes runoff and 48 percent becomes ET and canopy evaporation. Mean annual recharge for the southern aquifer sector, expressed as an equivalent depth uniformly spread over the sector area, is 30.6 in.

Compared to the north, recharge for the southern half of the island is more spatially variable owing to differences in underlying geology. Areas of local recharge maxima are underlain by limestone and include parts of (1) the older limestone cap, (2) the limestone plateau fringing the southeast coastline, (3) Orote Peninsula, and (4) internally drained areas. At these locations, runoff was assumed to be zero. Recharge in the older limestone cap, however, likely discharges at springs and into streams. Compared to surrounding grasslands and forested areas, recharge is greater at agriculture areas because of irrigation and lower rates of ET. Recharge is lowest in areas underlain by low-permeability volcanic soils because of high rates of runoff. Local recharge minima are in the drainage basins of the Pago and Ylig Rivers, which are predominantly underlain by Alutom volcanics, mostly covered by savanna complex, and have high rates of runoff.

The estimated fraction of total water inflow—rainfall, irrigation, septic-system leachate, and water-main leakage—that becomes groundwater recharge varies spatially (fig. 13). Most areas that receive water from septic systems and some areas that receive water from water-main leakage have a fraction greater than 70 percent. Recharge is between 40 and 60 percent of total water inflow at most areas that are underlain by limestone, except where stormwater runoff is routed to other areas. Areas with a low fraction, less than 30 percent, mostly occur in the southern half of the island where runoff from the volcanic land surface is relatively high. In parts of the volcanic uplands, recharge is less than 20 percent of total water inflow.

On a monthly basis, recharge follows the same trend as rainfall for northern Guam. Recharge is highest in August, the month with the most rainfall, and is lowest in March, the month with the least rainfall (fig. 14). During July through December, total mean monthly recharge is 60 percent of total mean monthly rainfall, and is 83 percent of the mean annual recharge. This result generally is consistent with the conclusion of Jones and Banner (2003), who indicated that recharge on northern Guam will typically only occur during July through November.

Comparison to Previous Water-Budget Studies

The water-budget model in this study differs from all previous water-budget investigations on Guam by directly accounting for canopy evaporation in forested areas, quantifying the evapotranspiration rate of each land-cover type, and accounting for evaporation from impervious areas. Recharge estimated in this study is greater than or similar to previous recharge estimates that used monthly water budgets, and is less than or similar to previous recharge estimates that used daily water budgets (table 5).

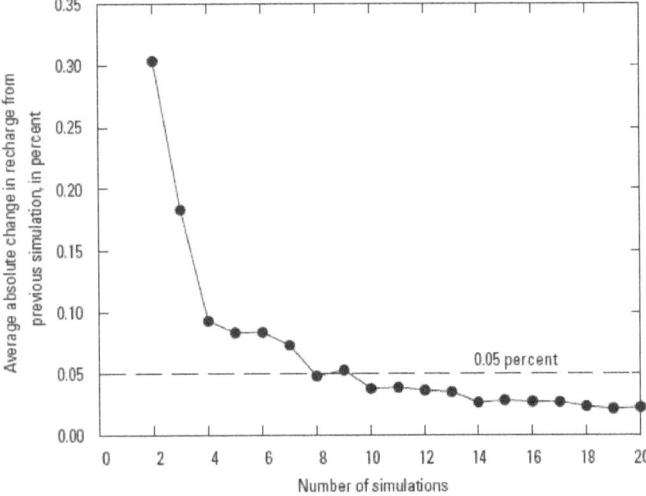

Figure 11. The average absolute percentage change in recharge of all the water-budget subareas with each successive model simulation.

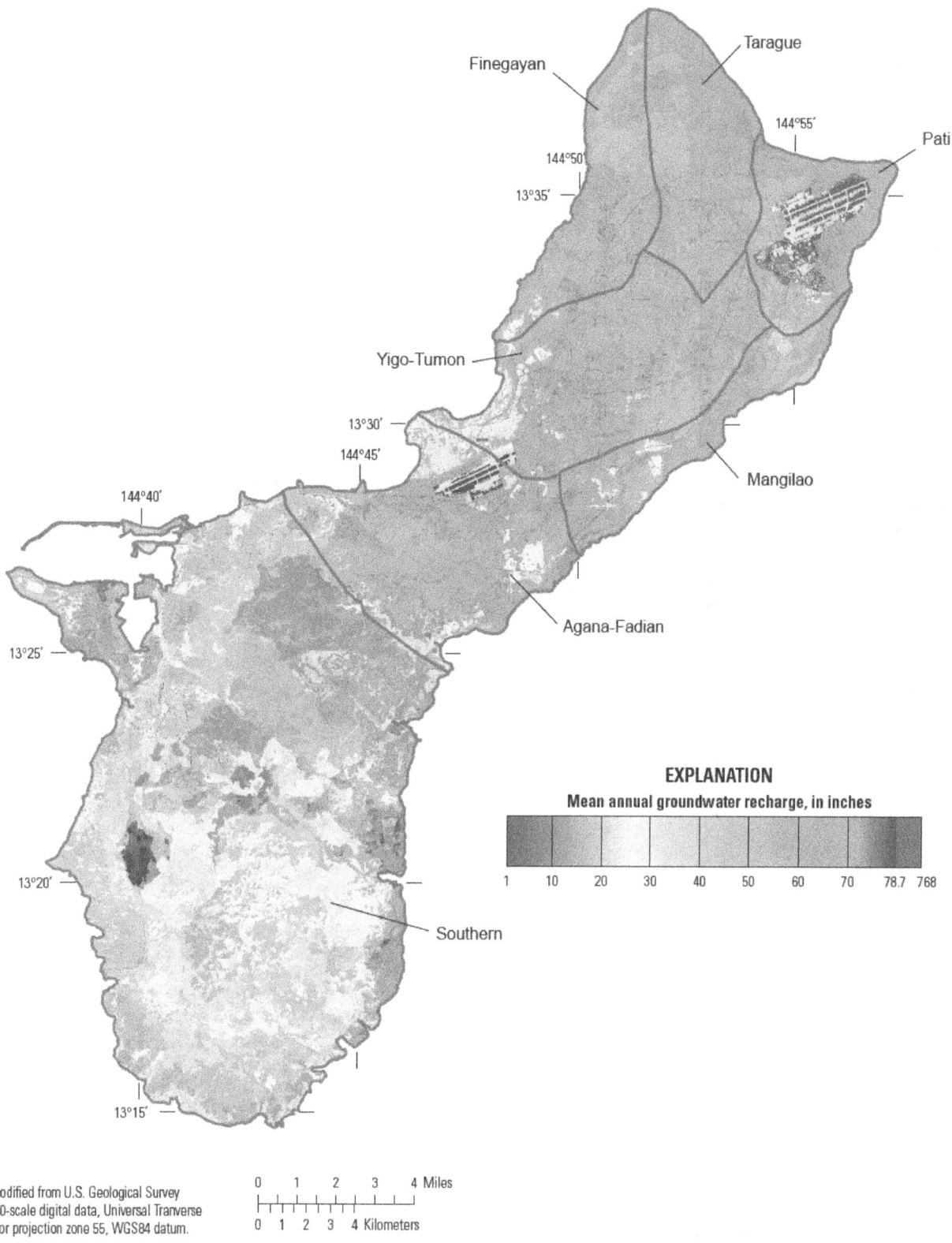

Figure 12. The distribution of mean annual groundwater recharge for baseline conditions on Guam calculated using the water-budget model. Subareas with artificially high recharge (colored purple) have mean annual recharge values greater than 78.7 inches because of water inflow from one or more of these components: water-main leakage, septic-system leachate, and stormwater runoff. Boundaries of aquifer sectors, as defined by Mink (1991), are shown in gray. Subareas with mean annual recharge greater than 78.7 inches are difficult to see at this scale.

Table 4. Mean annual water-budget estimates for baseline, drought, and possible future conditions for aquifer sectors on Guam.

[**Baseline** condition is 2004 land cover with mean rainfall during 1961–2005; **drought** condition is 2004 land cover with observed rainfall during 1969–1973; **future** condition is 2004 land cover with changes that would be induced by the military relocation to Guam with mean rainfall from 1961–2005; **future and drought** condition is 2004 land cover with changes that would be induced by the military relocation to Guam with observed rainfall during 1969–1973; mean monthly rainfall maps from Daly and Halblieb (2006b) were used to distribute rainfall for each condition; land-cover changes induced by military relocation to Guam are based on development projects in NAVFAC Pacific (2010c), Guam Waterworks Authority (2007c), and in an unpublished database provided by the Guam Bureau of Statistics and Plans; mi2, square miles; Mgal/d, million gallons per day; Irr, irrigation; Septic, septic-system leachate; Water main, water-main leakage; Total inflow, sum of rainfall, irrigation, septic-system leachate, and water-main leakage; ET, evapotranspiration exclusive of canopy evaporation; Canopy evap, canopy evaporation; Recharge expressed in in/yr (inches per year) is an equivalent depth of recharge uniformly spread over a sector area; water-budget components do not balance due to transfer of stormwater runoff across aquifer sector boundaries and rounding]

Aquifer sector(s)	Area (mi²)	Hypothetical condition	Water-budget estimate (Mgal/d)									Recharge (in/yr)
			Rain	Irr	Septic	Water main	Total inflow	Runoff	ET	Canopy evap	Recharge	
Tarague	15.37	Baseline	70.50	0.06	0.23	0.40	71.19	0.00	27.80	8.39	35.00	47.83
		Drought	56.61	0.10	0.23	0.40	57.34	0.00	26.30	7.42	23.62	32.28
		Future	70.50	0.03	0.23	0.40	71.16	0.00	27.88	8.23	35.06	47.91
		Future and drought	56.61	0.06	0.23	0.40	57.30	0.00	26.35	7.28	23.67	32.34
Finegayan	13.36	Baseline	61.98	0.02	0.12	0.68	62.80	0.00	25.53	6.96	30.30	47.63
		Drought	49.75	0.03	0.12	0.68	50.58	0.00	24.07	6.18	20.33	31.96
		Future	61.98	0.00	0.12	0.68	62.78	0.00	27.03	4.60	31.15	48.97
		Future and drought	49.75	0.00	0.12	0.68	50.55	0.00	25.43	4.07	21.04	33.08
Pati	12.04	Baseline	54.96	0.06	0.11	0.63	55.76	0.00	22.02	4.39	29.35	51.20
		Drought	44.14	0.10	0.11	0.63	44.98	0.00	20.77	3.88	20.33	35.46
		Future	54.96	0.06	0.11	0.63	55.76	0.00	22.03	4.36	29.36	51.22
		Future and drought	44.14	0.10	0.11	0.63	44.98	0.00	20.80	3.86	20.32	35.45
Yigo-Tumon	25.95	Baseline	119.72	0.14	1.37	2.58	123.81	0.68	52.70	8.51	61.97	50.16
		Drought	96.14	0.23	1.37	2.58	100.32	0.50	49.96	7.56	42.34	34.27
		Future	119.72	0.14	1.37	2.58	123.81	0.80	53.17	7.44	62.46	50.55
		Future and drought	96.14	0.22	1.37	2.58	100.31	0.58	50.42	6.61	42.75	34.60
Mangilao	10.54	Baseline	47.78	0.05	0.40	0.56	48.79	0.00	19.34	4.73	24.73	49.28
		Drought	38.38	0.08	0.40	0.56	39.42	0.00	18.25	4.19	16.98	33.84
		Future	47.78	0.05	0.40	0.56	48.79	0.00	19.65	3.94	25.20	50.22
		Future and drought	38.38	0.08	0.40	0.56	39.42	0.00	18.58	3.50	17.34	34.55

Table 4. Mean annual water-budget estimates for baseline, drought, and possible future conditions for aquifer sectors on Guam.—Continued

Aquifer sector(s)	Area (mi²)	Hypothetical condition	Water-budget estimate (Mgal/d)									Recharge (in/yr)
			Rain	Irr	Septic	Water main	Total inflow	Runoff	ET	Canopy evap	Recharge	
Agana-Fadian	25.04	Baseline	115.40	0.06	1.38	3.04	119.88	5.36	49.09	8.72	56.65	47.52
		Drought	92.67	0.10	1.38	3.04	97.19	4.19	46.26	7.74	38.97	32.69
		Future	115.40	0.06	1.38	3.04	119.88	5.42	49.15	8.20	57.07	47.87
		Future and drought	92.67	0.10	1.38	3.04	97.19	4.23	46.26	7.25	39.41	33.06
All northern	102.30	Baseline	470.34	0.39	3.61	7.89	482.23	6.04	196.48	41.70	238.00	48.86
		Drought	377.69	0.64	3.61	7.89	389.83	4.69	185.61	36.97	162.57	33.38
		Future	470.34	0.34	3.61	7.89	482.18	6.22	198.91	36.77	240.30	49.33
		Future and drought	377.69	0.56	3.61	7.89	389.75	4.81	187.84	32.57	164.53	33.78
Southern [1]	107.10	Baseline	528.65	0.39	1.22	4.90	535.16	123.61	222.05	33.42	156.07	30.61
		Drought	424.60	0.60	1.22	4.90	431.32	99.08	206.03	29.43	96.78	18.98
		Future	528.65	0.29	1.22	4.90	535.06	123.63	220.14	31.67	159.65	31.31
		Future and drought	424.60	0.45	1.22	4.90	431.17	99.10	204.50	27.92	99.69	19.55
Entire study area	209.40	Baseline	998.99	0.78	4.83	12.79	1,017.39	129.65	418.53	75.12	394.07	39.53
		Drought	802.30	1.24	4.83	12.79	821.15	103.77	391.64	66.40	259.35	26.01
		Future	998.99	0.63	4.83	12.79	1,017.24	129.85	419.05	68.44	399.95	40.11
		Future and drought	802.30	1.01	4.83	12.79	820.92	103.91	392.34	60.49	264.22	26.50

[1] Southern aquifer sector is the study area exclusive of the area encompassed by the northern aquifer sectors defined by Mink (1991).

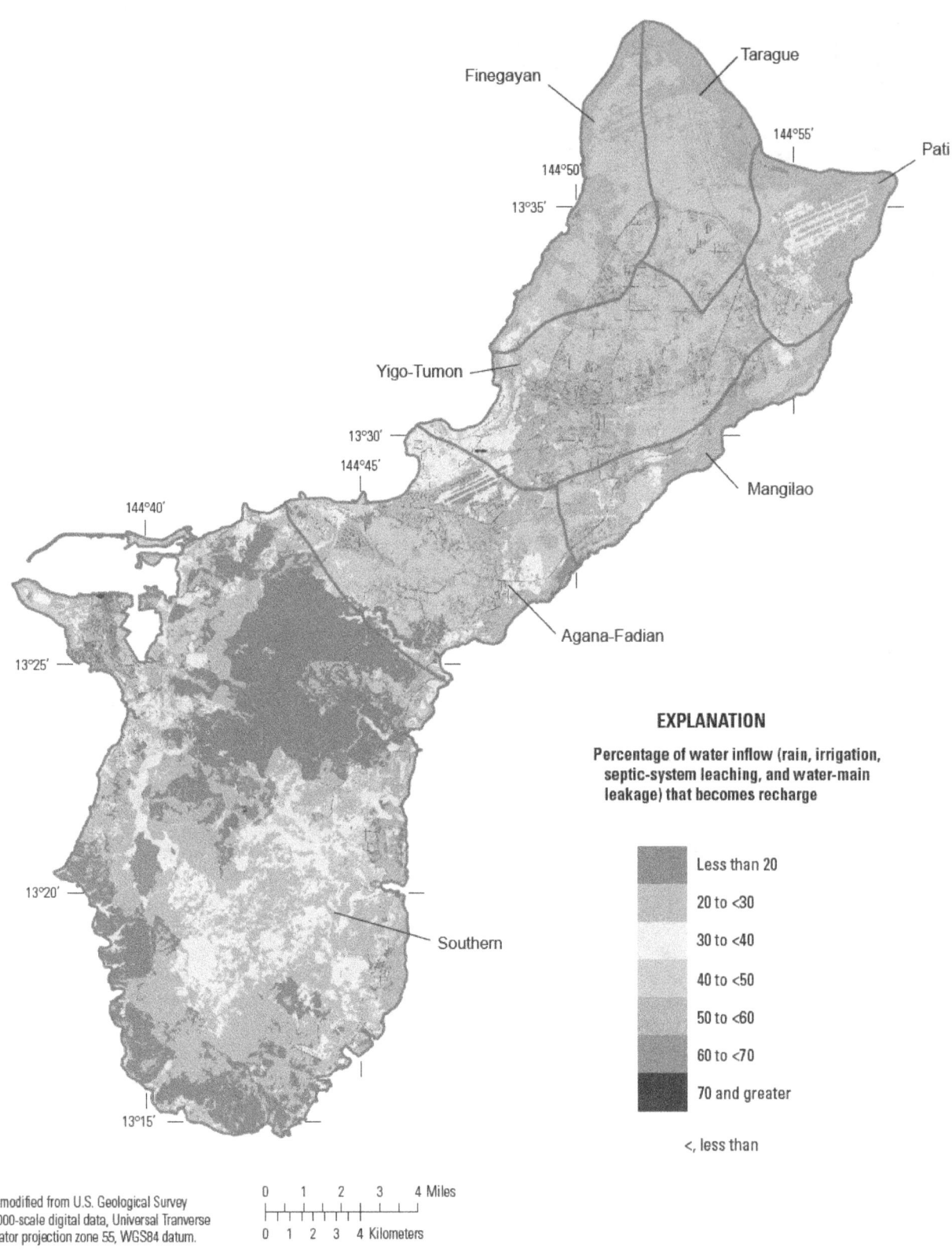

Figure 13. Estimated percentage of total water inflow that becomes groundwater recharge in the water-budget simulation for mean annual recharge for baseline conditions on Guam. Boundaries of aquifer sectors, as defined by Mink (1991), are shown in gray. Subareas with recharge that is at least 70 percent of water inflow are difficult to see at this scale.

For the entire water-budget area of Mink (1976), which spanned most of the northern half of the island, recharge estimated in this study is nearly twice as much as the "minimum" recharge estimate of Mink (1976) that assumed no runoff, and is about the same as the "probable" recharge estimate that assumed no runoff. The "minimum" recharge of Mink (1976) was estimated as mean monthly rainfall minus mean monthly pan evaporation, for months when mean rainfall was greater than pan evaporation; recharge was assumed to be zero for months when mean rainfall was less than pan evaporation. "Probable" recharge was estimated as mean annual rainfall minus annual ET. Annual ET in the north was estimated based on the difference between mean annual rainfall and mean annual runoff for two stream gages on the Ylig and Umatac Rivers in the south.

For each of the groundwater subbasins defined in Camp, Dresser & McKee Inc. (CDM) (1982), mean baseline recharge computed in this study is 32 to 49 percent greater than the 1982 CDM recharge estimates (table 6); total recharge to the combined area of all subbasins is 42 percent greater than the 1982 CDM recharge estimate. Compared to Camp, Dresser & McKee Inc. (1982) estimates, total water inflow to the aquifer subbasins estimated in this study is slightly greater, but total ET is 15 percent lower. Similar to Mink (1976), Camp, Dresser & McKee Inc. (1982) estimated recharge as mean monthly rainfall minus mean monthly ET. Mean monthly ET was estimated from mean monthly temperature and percent daytime hours using the Blaney-Criddle method (Blaney and Criddle, 1950).

Mink (1991) extended the Camp, Dresser & McKee Inc. (1982) groundwater-subbasin boundaries to the coastline. The revised groundwater-subbasin areas, which encompass most of the northern half of the island, were reclassified as "aquifer sectors" (fig. 12). Some aquifer sectors retain the Camp, Dresser & McKee Inc. (1982) subbasin names, while other aquifer sectors have slightly different names. For each aquifer sector, Mink (1991) reported sustainable-yield estimates; recharge, however, was reported only for the total area of all aquifer sectors. For the total area of all aquifer sectors, recharge computed in this study is about 6 percent less than the recharge reported by Mink (1991) (table 5). Compared to Mink (1991), mean rainfall for this study is 11 percent greater; total ET is 41 percent greater. Mink (1991) estimated recharge as the difference between mean monthly rainfall and ET, where ET was assumed to be 73 percent of rainfall from January to May, and 3.3 in. from June to December. The studies in which these ET assumptions were based on were not referenced in Mink's report.

For parts of the Yigo-Tumon and Finegayan aquifer sectors, recharge estimated by Jocson and others (2002) is 26 percent higher than recharge estimated in this report. Annual recharge estimated by Jocson and others (2002) was 1.6 m (63 inches), which, when applied to the study area (36.9 mi^2) equates to 110 Mgal/d. Jocson and others (2002) assumed potential ET for the study area was equal to pan evaporation measurements from weather station 914229, regardless of land cover. Daily recharge equaled daily rainfall minus daily pan evaporation for days when rainfall exceeded pan evaporation; recharge was assumed to be zero on days when rainfall was less than pan evaporation. Although not stated in the report, this method assumes either that (1) on a given day, no water is stored in the soil at the end of the day, so it must be lost to ET or become recharge, or (2) any water stored in the soil at the end of the day is not subject to ET on any subsequent days, and it will ultimately become recharge. Thus, on days when rainfall was less than pan evaporation, actual ET was equal to rainfall and was less than pan evaporation; on days when rainfall was zero, actual ET was zero. The method used by this study to estimate ET is different than the method used by Jocson and others (2002). Because of this, and by directly accounting for canopy evaporation, ET estimated for this study is 46 percent higher than ET estimated by Jocson and others (2002).

For parts of the Yigo-Tumon and Mangilao aquifer sectors, recharge estimates by Habana and others (2009) are between 6 and 34 percent higher than recharge estimated in this study. Similar to Jocson and others (2002), Habana and others (2009) assumed potential ET for the study area was equal to pan evaporation, regardless of land cover. However, unlike Jocson and others (2002), but similar to this study, GIS soil maps and associated available water capacities were incorporated into the water budget, and the actual rate of ET for a given day was estimated on the basis of soil-moisture content. Habana and others (2009) used three soil-moisture models to derive three recharge estimates. For soil-moisture model 1, the rate of evapotranspiration was assumed to decrease linearly as soil moisture is reduced from field capacity to the wilting point. For soil-moisture model

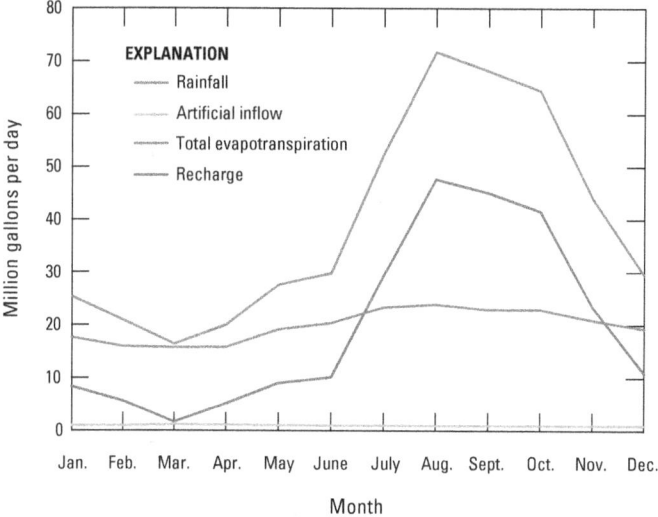

Figure 14. Estimated mean monthly rainfall, total evapotranspiration, artificial inflows (irrigation, water-main leakage, and septic-system leachate), and recharge, 1961–2005, for the northern aquifer sectors of Guam defined by Mink (1991).

Table 5. Comparison of the baseline water budget calculated in this study with previous studies for selected areas of northern Guam.

[Unless noted, the area of each study was estimated using a Geographic Information System; The water-budget values from Mink (1991) were converted from inches per year to million gallons per day based on the reported study area; The water-budget values from Jocson and others (2002) and Habana and others (2009) were converted from inches per year to million gallons per day based on estimated study area. Baseline condition is 2004 land cover with mean rainfall during 1961–2005 distributed according to Daly and Halbleib (2006b); mi², square miles; Mgal/d, million gallons per day; Irr, irrigation; Septic, septic-system leachate; Water main, water-main leakage; Total inflow, sum of rainfall, irrigation, septic-system leachate and water-main leakage; Total ET, evapotranspiration inclusive of canopy evaporation; –, value not given in referenced study; SM, soil-moisture; water-budget components do not balance due to transfer of stormwater runoff across study-area boundaries and rounding]

Study location	Study	Time step	Area (mi²)	Water-budget estimate (Mgal/d)							
				Rain	Irr	Septic	Water main	Total inflow	Runoff	Total ET	Recharge
Most of northern Guam	Mink (1976), minimum [a]	Monthly	[b]94.6	405.1	–	–	–	405.1	0.0	[c]275.9	129.2
	Mink (1976), minimum [d]	Monthly	[b]94.6	405.1	–	–	–	405.1	20.3	275.9	109.1
	Mink (1976), probable [a]	Monthly	[b]94.6	405.1	–	–	–	405.1	0.0	[c]173.1	232.0
	Mink (1976), probable [d]	Monthly	[b]94.6	405.1	–	–	–	405.1	20.3	173.1	211.7
	This study, baseline	Daily	98.78	454.14	0.39	3.52	7.42	465.47	3.83	230.39	231.26
Northern aquifer subbasins	Camp, Dresser & McKee Inc. (1982)	Monthly	[b]67.9	304.63	–	–	–	304.63	0.0	192.72	111.91
	This study, baseline	Daily	69.10	318.34	0.31	2.89	5.46	327.00	3.29	164.20	159.08
Northern aquifer sectors	Mink (1991)	Monthly	[b]100.3	422.9	–	–	–	422.9	0.0	168.5	254.4
	This study, baseline	Daily	102.30	470.34	0.39	3.61	7.89	482.23	6.04	238.18	238.00
Parts of Yigo-Tumon and Finegayan aquifer sectors	Jocson and others (2002)	Daily	36.86	170.0	–	–	–	170.0	0.0	[c]60.0	110.0
	This study, baseline	Daily	36.86	170.66	0.19	1.55	3.45	175.85	0.92	87.63	87.28
	This study, baseline [e]	Daily	36.86	161.93	0.22	1.55	3.45	167.15	0.85	87.09	79.20
Parts of Yigo-Tumon and Mangilao aquifer sectors	Habana and others (2009), SM-model 1	Daily	18.02	87.73	–	–	–	87.73	0.0	30.03	57.70
	Habana and others (2009), SM-model 2	Daily	18.02	87.73	–	–	–	87.73	0.0	42.08	45.67
	Habana and others (2009), SM-model 3	Daily	18.02	87.73	–	–	–	87.73	0.0	39.60	48.15
	This study, baseline	Daily	18.02	82.79	0.07	0.97	1.86	85.69	0.30	41.81	43.11
	This study, baseline [e]	Daily	18.02	78.56	0.08	0.97	1.86	81.47	0.27	41.59	39.17

[a] Assuming no runoff.

[b] Area reported in the study.

[c] ET not reported in study, assumed to equal rainfall minus recharge.

[d] Assuming runoff is 5 percent of rainfall.

[e] 2004 land cover and observed rainfall and reference evapotranspiration during 1982–95.

Table 6. Comparison of the baseline water budget calculated in this study with Camp, Dresser & McKee Inc. (1982).

[Unless noted, the area of each study was estimated using a Geographic Information System; CDM, Camp, Dresser & McKee Inc.; Baseline condition is 2004 land cover with mean rainfall during 1961–2005 distributed according to Daly and Halbleib (2006b); mi², square miles; Mgal/d, million gallons per day; Irr, irrigation; Septic, septic-system leachate; Water main, water-main leakage; Total inflow, sum of rainfall, irrigation, septic-system leachate and water-main leakage; Total ET, evapotranspiration inclusive of canopy evaporation; –, value not given in referenced study; water-budget components do not balance due to transfer of stormwater runoff across subbasin boundaries and rounding]

Subbasin	Study	Area (mi²)	Water-budget estimate (Mgal/d)							
			Rain	Irr	Septic	Water main	Total inflow	Runoff	Total ET	Recharge
Andersen	CDM (1982)	[a]6.83	31.41	–	–	–	31.41	0.00	20.26	11.15
	This study, baseline	7.07	32.19	0.02	0.11	0.46	32.78	0.00	16.04	16.64
Agafa Gumas	CDM (1982)	[a]11.68	51.57	–	–	–	51.57	0.00	31.92	19.65
	This study, baseline	11.84	54.46	0.06	0.24	0.38	55.14	0.00	28.66	26.48
Agana	CDM (1982)	[a]16.13	73.12	–	–	–	73.12	0.00	47.67	25.45
	This study, baseline	16.49	76.18	0.07	1.01	1.58	78.84	3.29	39.04	36.45
Finegayan	CDM (1982)	[a]7.61	33.55	–	–	–	33.55	0.00	20.63	12.92
	This study, baseline	7.39	34.46	0.02	0.18	0.69	35.35	0.00	18.34	17.00
Mangilao	CDM (1982)	[a]4.44	19.84	–	–	–	19.84	0.00	12.37	7.47
	This study, baseline	4.45	20.23	0.00	0.17	0.37	20.77	0.00	10.50	10.27
Yigo	CDM (1982)	[a]21.23	95.14	–	–	–	95.14	0.00	59.87	35.27
	This study, baseline	21.86	100.82	0.14	1.18	1.98	104.12	0.00	51.62	52.24
All subbasins	CDM (1982)	[a]67.92	304.63	–	–	–	304.63	0.00	192.72	111.91
	This study, baseline	69.10	318.34	0.31	2.89	5.46	327.00	3.29	164.20	159.08

[a] Area reported in the study.

2, the ET rate was assumed to decrease exponentially as soil moisture is reduced from field capacity to the wilting point. For soil-moisture model 3, ET occurs at the potential rate as the soil-moisture is reduced from the field capacity to an intermediate soil-moisture value greater than the wilting point; the ET rate decreases exponentially as soil moisture is reduced from the intermediate soil-moisture value to the wilting point. The soil-moisture model employed in this study is from Allen and others (1998) and is most similar to model 2 of Habana and others (2009). Compared to most other soil-moisture models used in water budgets, model 1, presumably derived from Thornthwaite and Mather (1955), is considered an extreme model because it will estimate a relatively low ET and will produce a relatively high recharge estimate. In contrast, models 2, 3, and the model of Allen and others (1998) will produce relatively low recharge estimates.

The recharge estimates of Jocson and others (2002) and Habana and others (2009) used daily rainfall and pan evaporation data from 1982 to 1995. Therefore, to aid comparisons with these previous studies, the water-budget model developed for this current study was run using observed annual rainfall and estimated reference ET during 1982–95; all other baseline model inputs were kept the same. Compared to baseline recharge and the recharge estimates from these previous studies, mean annual recharge for the study areas during 1982–95 was lower (table 5).

Drought Conditions

Analysis of the effect of drought conditions on Guam was based on annual rainfall at weather station 914226 at the Guam International Airport. Rainfall during 1969–73, the period with the lowest 5-year rainfall average between 1961 and 2005 at this station, was used to represent drought conditions. The water-budget model was run using the same input as the baseline simulation, except that rainfall from the historical drought period was used instead of mean rainfall. The water-budget model was run for 40 5-yr simulations, and the results were averaged.

For the entire island, recharge during drought conditions was 259.3 Mgal/d and thus was 34 percent lower than recharge computed for baseline conditions (table 4). Estimated mean annual rainfall during this period was about 20 percent below the mean for 1961–2005. Of the total water inflow, 13 percent becomes runoff, and 56 percent becomes ET and canopy evaporation.

The relative distribution of groundwater recharge for drought conditions is similar to the baseline distribution; however, the overall rates of recharge are lower because of lower rainfall (fig. 15). For the northern aquifer sectors, mean annual recharge is 162.6 Mgal/d during drought conditions, which is 32 percent lower than mean baseline recharge. Expressed as a uniform depth spread over each northern aquifer sector

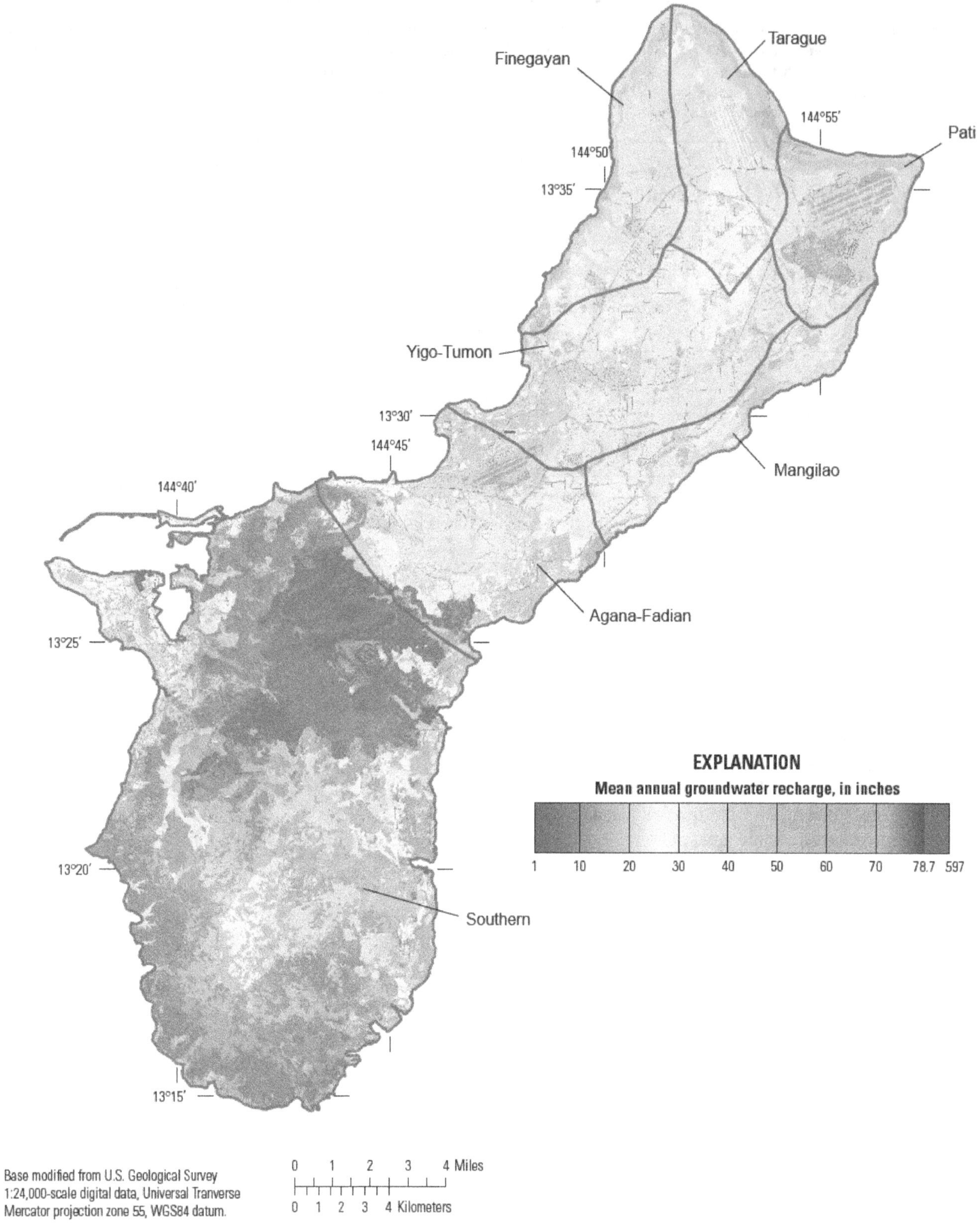
Finegayan
Tarague
144°55′
Pati
144°50′
13°35′
Yigo-Tumon
13°30′
144°45′
144°40′
Mangilao
Agana-Fadian
13°25′
EXPLANATION
Mean annual groundwater recharge, in inches
1
10
20
30
40
50
60
70
78.7
597
13°20′
Southern
13°15′
Base modified from U.S. Geological Survey 1:24,000-scale digital data, Universal Tranverse Mercator projection zone 55, WGS84 datum.
0 1 2 3 4 Miles
0 1 2 3 4 Kilometers

Figure 15. The distribution of mean annual groundwater recharge for drought conditions on Guam calculated using the water-budget model. Subareas with artificially high recharge (colored purple) have mean annual recharge values greater than 78.7 inches because of water inflow from one or more of these components: water-main leakage, septic-system leachate, and stormwater runoff. Boundaries of aquifer sectors, as defined by Mink (1991), are shown in gray. Subareas with mean annual recharge greater than 78.7 inches are difficult to see at this scale.

area, mean annual recharge is between 32.0 and 35.5 in. per sector. Mean annual recharge for the southern aquifer sector is 19.0 in. during drought conditions. At many subareas in the steeply dissected volcanic uplands, mean annual recharge is less than 10 in. Subareas classified as savanna complex were most affected by drought conditions compared to other land covers, with recharge averaging 48 percent lower than baseline recharge. For all other land-cover classes, except for water bodies and areas receiving septic-system leachate, recharge during drought conditions was between 27 and 39 percent lower than baseline recharge.

Future Land Cover

Groundwater recharge was also estimated for a future land-cover scenario that represents potential land-cover conditions on Guam following the proposed military relocation (NAVFAC Pacific, 2010a). This scenario incorporates construction and development projects directly associated with the military relocation that would alter the current land cover including those related to (1) the main cantonment, family housing, training, airfield, and waterfront actions (NAVFAC Pacific, 2010c); and (2) road improvement projects (NAVFAC Pacific, 2010b). With the exception of road improvement projects, potential developments within the private and public sector needed to accommodate the population growth incurred during the military relocation are not presented in the FEIS. Therefore, potential private sector and public developments in the future land-cover scenario were obtained from (1) an unpublished GIS database of approved future development projects supplied by the Guam Bureau of Statistics and Plans (Victor Torres, written commun., 2010), and (2) potential development projects listed in the 2005, 2010, and 2020 land-use forecasts within the Guam Waterworks Authority (2007c) Water Resource Master Plan. Although "The Draft North and Central Guam Land Use Plan" (ICF International, 2009) serves as a guide for future development within the northern half of Guam, the map classification units of its future land-use map are too general to accurately assign quantifiable attributes required for the water budget; therefore, it was not used in this study.

Potential land-cover changes incorporated into the future land-cover scenario were assigned to a potential land-cover class and were compiled into a future land-cover map (fig. 16). At areas where no land-cover changes are expected, the future land cover was assumed to be the same as the 2004 land-cover map (U.S. Department of Agriculture, 2006a). Land-cover parameters of the potential land-cover classes were estimated from 2004 land-cover parameters (table 2). Construction projects related to the military relocation were assigned to urban builtup or urban cultivated land covers on the basis of project descriptions within the FEIS (NAVFAC Pacific, 2010c). The main cantonment was divided into two land covers, main cantonment I and II. The pervious fraction for main cantonment I and II, 0.73 and 0.23, respectively, was estimated in a GIS using maps from the FEIS (NAVFAC Pacific, 2010c).

All other land-cover parameters for main cantonment I and II were the same as the urban builtup class. New housing developments and subdivision projects from the Guam Bureau of Statistics and Plans database and the GWA Water Resource Master Plan (Guam Waterworks Authority, 2007c) were assigned to the housing development land-cover class. The housing development land-cover class was assigned a pervious fraction of 0.54, a crop coefficient of 1.10, a root depth of 12 in., and a depletion fraction of 0.47. The land-cover parameters for the housing development class were estimated from existing housing developments in the 2004 land-cover map, which are about one-third urban cultivated and two-thirds urban builtup. The family housing land-cover class, derived from the FEIS (NAVFAC Pacific, 2010c), was assigned the same land-cover parameters that were assigned to the housing development class. Development projects such as hotels, apartments, and parking lots from the Guam Bureau of Statistics and Plans database and the GWA Water Resource Master Plan (Guam Waterworks Authority, 2007c) were assigned to the urban builtup class. Although it is not known which of the proposed development projects from these sources will be built, the projects that were included in this scenario represent an approximation of the potential magnitude of land-cover changes that would occur on Guam in order to accommodate the military relocation and population growth.

For the future land-cover scenario, estimated recharge for Guam is 399.9 Mgal/d, which is about 1 percent greater than the baseline recharge estimate that was based on 2004 land cover. For this scenario, all water-budget parameters are the same as baseline conditions, except the future land-cover map was used in place of the 2004 land-cover map. Estimated recharge for the northern aquifer sectors was less than 1 percent greater than the baseline recharge estimate. Here, the slight increase in ET was offset by a proportionally greater decrease in canopy evaporation. Estimated recharge for the southern aquifer sector was about 2 percent greater than the baseline recharge estimate, owing to slight decreases in ET and canopy evaporation. Localized spatial variations in recharge, such as relatively low recharge areas in paved areas and relatively high recharge rates in nearby grassy areas, are not fully resolved in the water-budget model estimate; however, the total recharge rates area representative of the overall developments.

Drought Conditions and Future Land Cover

The water-budget model was also run for a scenario that used the future land-cover map and rainfall during drought conditions; all other water-budget parameters were kept the same as baseline conditions. The water-budget model was run for 40 5-yr simulations, and the results were averaged. For this scenario, estimated recharge on Guam is 264.2 Mgal/d, which is about 33 percent lower than the baseline recharge estimate. Estimated recharge for the northern aquifer sectors on Guam is 164.5 Mgal/d, which is about 31 percent lower than the baseline recharge estimate. Estimated recharge for

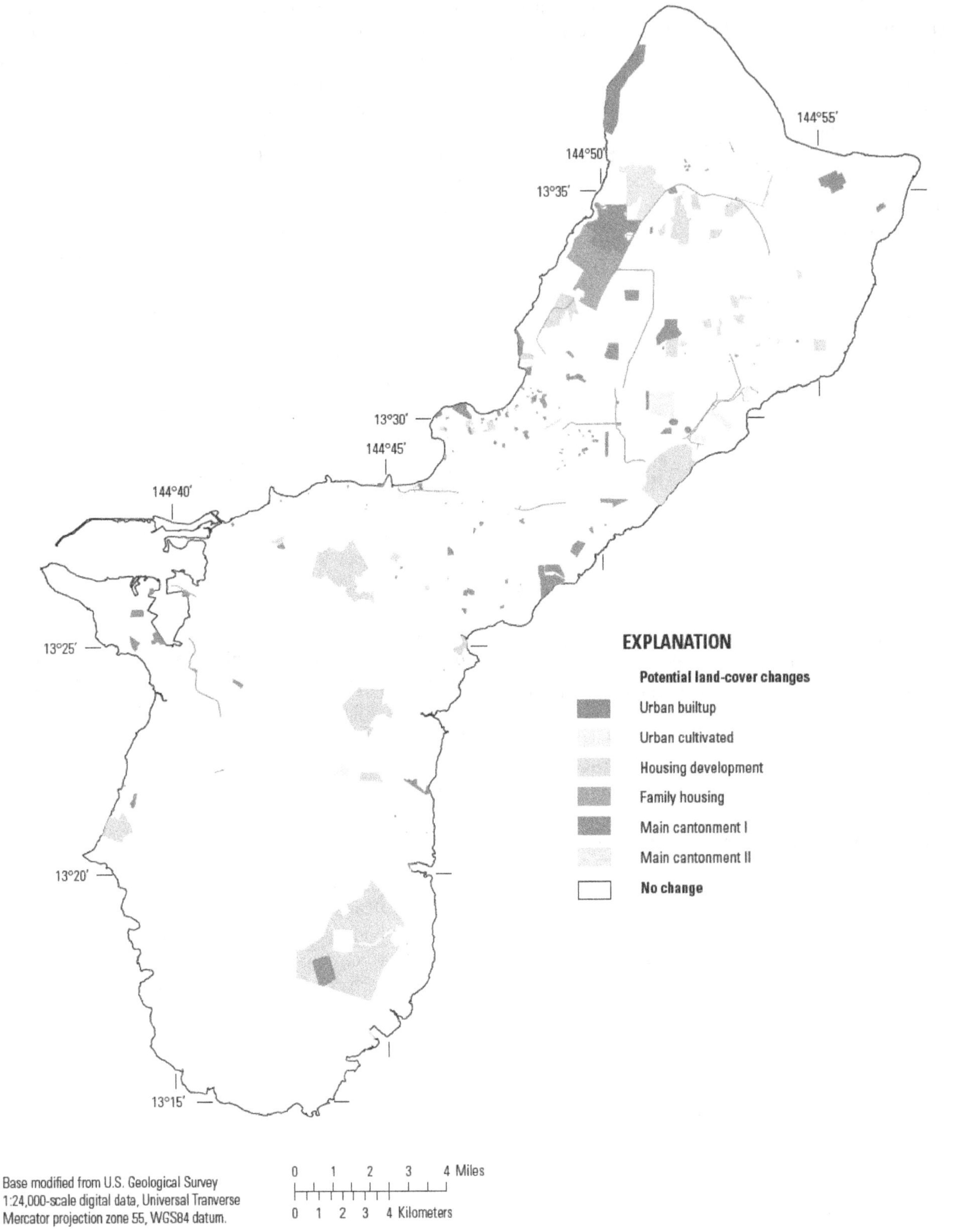

Figure 16. Potential land-cover changes used in the future land-cover scenario water-budget recharge estimate for Guam. Potential land-cover changes are due to a proposed military relocation to Guam and population growth and are based on development projects in the Final Environmental Impact Statement (NAVFAC Pacific, 2010c), Guam Waterworks Authority (2007c), and an unpublished database from the Guam Bureau of Statistics and Plans.

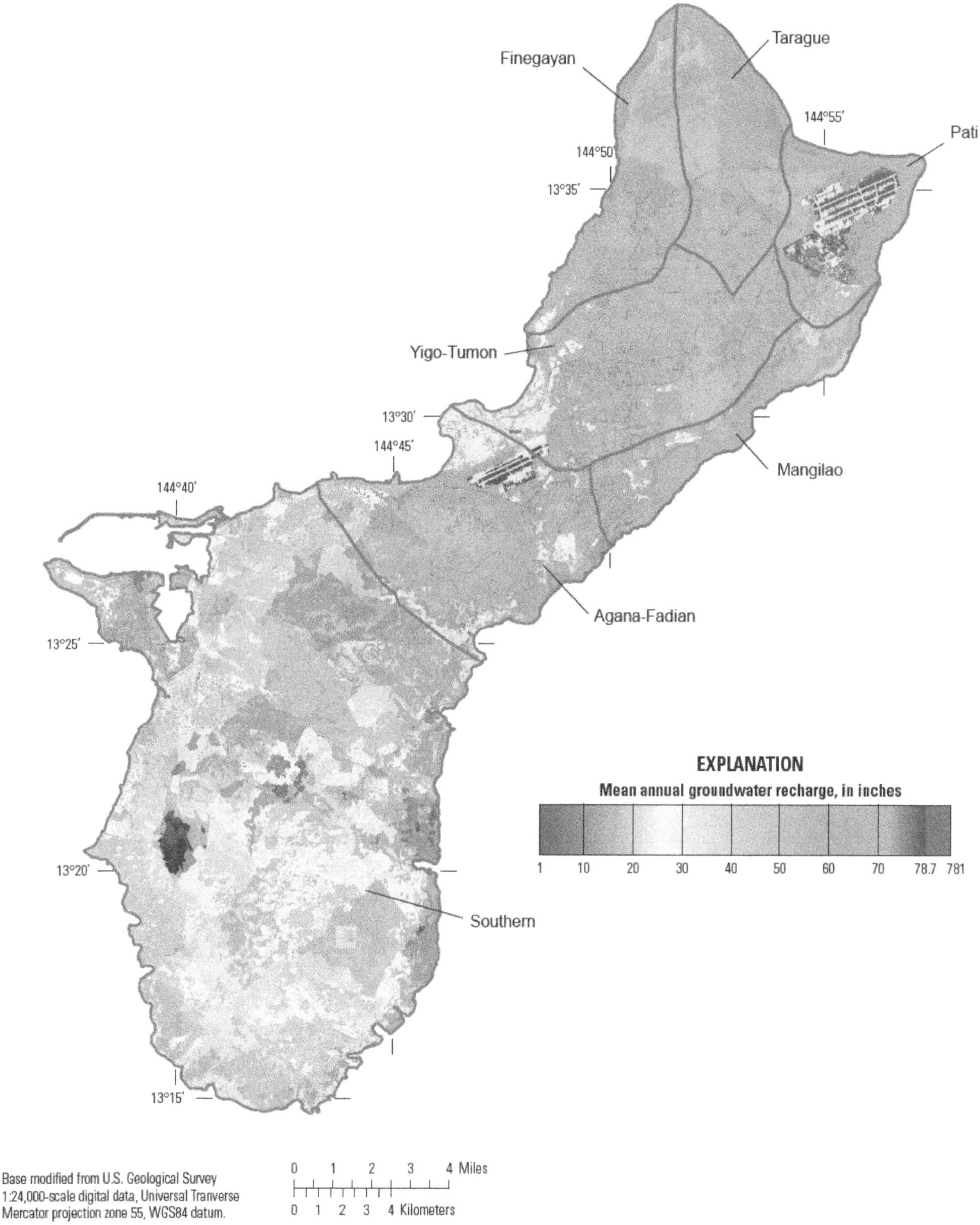

Figure 17. The distribution of mean annual groundwater recharge for future land-cover conditions on Guam using the water-budget model. Subareas with artificially high recharge (colored purple) have mean annual recharge values greater than 78.7 inches because of water inflow from one or more of these components: water-main leakage, septic-system leachate, and stormwater runoff. Boundaries of aquifer sectors, as defined by Mink (1991), are shown in gray. Subareas with mean annual recharge greater than 78.7 inches are difficult to see at this scale.

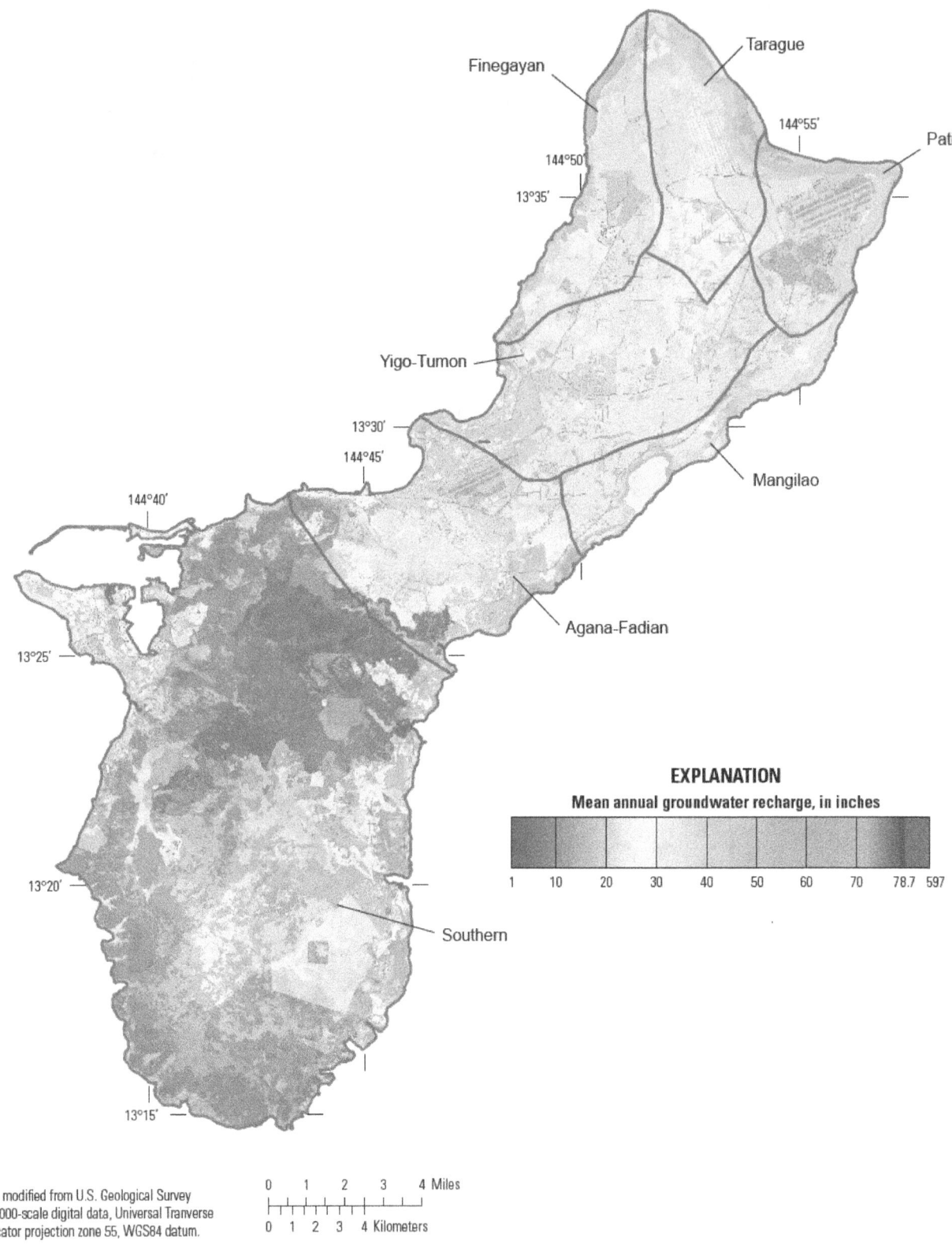

Figure 18. The distribution of mean annual groundwater recharge for future land-cover and drought conditions on Guam using the water-budget model. Subareas with artificially high recharge (colored purple) have mean annual recharge values greater than 78.7 inches because of water inflow from one or more of these components: water-main leakage, septic-system leachate, and stormwater runoff. Boundaries of aquifer sectors, as defined by Mink (1991), are shown in gray. Subareas with mean annual recharge greater than 78.7 inches are difficult to see at this scale.

the southern aquifer sector on Guam is 99.7 Mgal/d, which is about 36 percent lower than the baseline recharge estimate.

Sensitivity Analysis

Uncertainty exists in many of the water-budget inputs used in this study. The values used in the water-budget model were deemed to be those most reasonable. To analyze the effect that uncertainty in water-budget inputs has on estimated recharge, the water budget was rerun while changing one input value at a time within a reasonable range. A range of recharges for each test was computed by holding all other inputs at their original values and varying the test input. The resulting recharge estimates for the entire island, northern aquifer sectors, and southern aquifer sector are compared to baseline recharge (table 7). The parameters tested were (1) available water capacity, (2) crop coefficient, (3) fraction of stormwater runoff captured by storm-drain systems, (4) root depth, (5) ratio of runoff to rainfall, (6) canopy storage capacity, (7) the net-precipitation rate in forests, and (8) water-main leakage. For available water capacity, the range chosen for testing was between the high and low values published in U.S. Department of Agriculture (2009). For crop coefficients and the fraction of stormwater runoff captured by storm drains, baseline values were increased by 20 percent and decreased by 20 percent. For root depth and ratio of runoff to rainfall, baseline values were increased by 50 percent and decreased by 50 percent. For the ratio of runoff to rainfall, baseline values were also increased by 5 percentage units because much of the northern half of the island was assumed to have zero runoff for baseline conditions. For canopy storage capacity of forests, baseline values were increased by 40 percent and decreased by 40 percent. For the net-precipitation rate in forests, which is measured as a percentage of rainfall, baseline values were increased by 15 percentage units and decreased by 15 percentage units. Based on the range of UFW values reported in FEIS (NAVFAC Pacific, 2010d), recharge was estimated for water-main leakage that was (1) 50 percent of current input to each water-main system, (2) 15 percent of current input to each water-main system, and (3) zero for each water-main system.

Parameters with minor effects on recharge were (1) available water capacity, (2) fraction of stormwater runoff captured by storm-drain systems, (3) root depth, (4) canopy storage capacity, and (5) water-main leakage (table 7). Varying these parameters within the ranges listed in table 7 resulted in a net difference in recharge of less than 7 percent relative to the baseline scenario for each aquifer sector.

Parameters with relatively larger effects on recharge were the ratio of runoff to rainfall, net precipitation rates, and crop coefficients. In the northern aquifer sectors, for the case in which runoff was assumed to be 5 percent of rainfall, recharge was about 9 percent less than baseline conditions; for the case in which net precipitation was assumed to be 100 percent of rainfall (canopy evaporation was assumed to be zero), recharge increased by about 10 percent.

Chloride Mass-Balance Method

The chloride mass-balance method was used as a second, independent method of estimating recharge on Guam. The chloride mass-balance method is based on the assumption that bulk (wet and dry) deposition is the only source of chloride in groundwater and surface-water runoff. In areas where surface runoff is negligible, the chloride flux to the land surface equals the chloride flux to the water table, under steady-state conditions. Chloride is deposited on the land surface in rainfall and dry deposition of aerosols. On oceanic islands such as Guam, ocean spray from the sea surface is carried and deposited onto the island by wind and rain. Some of the rain that falls on the land surface evaporates or is transpired by plants. The flux of chloride back to the atmosphere due to these processes, however, is assumed to be zero. Consequently the concentration of chloride in the remaining soil water increases. At depths below the root zone, chloride in the remaining water is assumed to be constant (Scanlon, 1991). Therefore, the chloride concentration in the remaining water can be used to estimate the fraction of rain water that recharges the water table.

Atmospheric deposition of chloride was measured using bulk-deposition collectors. The fraction of rainfall that percolated past the root zone was determined from spring samples and groundwater samples. At the groundwater sampling locations for this study, human sources such as septic systems contribute minimal amounts of chloride to the water, and saltwater intrusion is nonexistent or assumed negligible. One exception is at well Y-15, where there could potentially be chloride input from septic systems and water-main leakage. Chloride input from seawater at well Y-15 is unlikely, however, because drilling records indicate that the bottom of this well is about 78 ft above mean sea level. A mass balance of chloride in bulk deposition, surface runoff, and groundwater is expressed in the following equation (Prych, 1995; Maurer and others, 1996):

$$P \times C_p = (GWR \times C_g) + (SWR \times C_p) \qquad (17)$$

where

P	=	annual precipitation [inches],
C_p	=	concentration of chloride in bulk deposition [milligrams per liter],
GWR	=	annual groundwater recharge [inches],
C_g	=	concentration of chloride in groundwater [milligrams per liter],
SWR	=	annual surface-water runoff [inches],

Rearranging the terms in equation 17 and solving for GWR gives:

$$GWR = \frac{(P \times C_p) - (SWR \times C_p)}{C_g} \qquad (18)$$

Table 7. Results of sensitivity testing for selected parameters used in the water-budget model for Guam.

[See figure 12 for locations of aquifer sectors; %, percent]

Parameter(s)	Adjusted parameter value	Percent difference in recharge relative to baseline conditions								
		Entire study area	All northern aquifer sectors	Southern	Tarague	Agana-Fadian	Pati	Finegayan	Mangilao	Yigo-Tumon
Available water capacity	Low reported value[1]	1.8	1.5	2.2	1.8	1.3	1.8	1.5	1.8	1.4
	High reported value[1]	-1.5	-1.2	-1.8	-1.4	-1.1	-1.5	-1.2	-1.5	-1.1
Crop coefficient	120% of baseline	-10.1	-8.3	-12.7	-8.7	-8.2	-7.0	-9.2	-8.2	-8.5
	80% of baseline	12.6	10.1	16.7	10.6	9.9	8.5	11.3	9.9	10.3
Fraction of stormwater runoff captured by storm drains	20 percentage units higher than baseline	-0.3	-0.3	-0.2	0.0	-1.1	0.1	0.0	0.0	-0.4
	20 percentage units lower than baseline	0.3	0.4	0.2	0.0	1.1	-0.1	0.0	0.0	0.4
Root depth	150% of baseline	-1.1	-0.4	-2.0	-0.2	-1.2	-0.2	-0.1	-0.2	-0.2
	50% of baseline	4.2	2.5	6.7	1.6	3.9	2.6	1.4	1.6	2.5
Ratio of runoff to rainfall	5 percentage units higher than baseline	-10.4	-8.6	-13.2	-8.9	-8.5	-8.3	-8.9	-8.5	-8.4
	150% of baseline	-11.8	-0.6	-28.9	0.0	-2.4	0.0	0.0	0.0	0.0
	50% of baseline	13.5	0.7	33.0	-0.1	2.8	-0.1	0.1	-0.1	0.0
Canopy storage capacity	140% of baseline	0.3	0.1	0.6	0.2	0.1	0.3	0.0	0.2	0.0
	60% of baseline	-0.4	-0.1	-0.7	-0.2	-0.2	-0.3	0.0	-0.3	-0.1
Net-precipitation rates in forest (as a percentage of rainfall)	15 percentage units higher than baseline	11.3	10.3	12.8	14.8	8.0	8.2	14.3	10.6	8.7
	15 percentage units lower than baseline	-15.2	-13.6	-17.7	-18.9	-11.6	-12.0	-17.2	-15.1	-10.8
Water-main leakage	50% of input to each system	3.1	3.0	3.1	0.9	5.4	0.5	2.1	2.2	4.0
	15% of input to each system	-1.3	-1.4	-1.2	-0.5	-2.1	-1.4	-0.9	-0.9	-1.7
	No leakage	-3.2	-3.3	-3.1	-1.1	-5.4	-2.1	-2.2	-2.3	-4.2

[1]High and low values reported in U.S. Department of Agriculture (2009).

For the bulk-deposition stations in northern Guam, where runoff is assumed to be negligible, equation 18 simplifies to:

$$GWR = \frac{(P \times C_p)}{C_g} \qquad (19)$$

Implicit in the derivation and uses of equations 18 and 19 is the assumption of "plug flow," or piston flow, which assumes that (1) the direction of water flow and chloride transport is vertical and downward, (2) areal distributions of the rate of percolation of water and of chloride on the local scale (a few tenths of a meter) are uniform (no preferred pathways), (3) all chloride is dissolved in soil water, and the distribution of the dissolved chloride in the soil water is relatively uniform within a pore (no solid chloride phase, sorption by soil, or anion exclusion), and (4) advection is the dominant mode of chloride transport, and diffusion is relatively unimportant. Additional assumptions are that (5) minerals in the soil are not a source of chloride, and the only sources are precipitation and dry-atmospheric deposition, (6) seasonal variations in chloride concentration in groundwater are small, and (7) the concentration of chloride in surface-water runoff is the same as that in precipitation. The method is still valid if chloride is taken up by growing vegetation as long as it is also released by decaying vegetation at the same rate.

Data Collection

Atmospheric deposition of chloride at the land surface was measured at five bulk-deposition stations: four on the northern half of the island and one on the southern half near Fena Valley Reservoir (fig. 3). Three of these stations measured atmospheric deposition between March 2010 and May 2011; the other two stations measured deposition for only part of this period (table 8). Bulk-deposition sampling methods were based on Scholl and Ingebritsen (1995), Scholl and others (1995), and Scholl and others (1996). Each bulk-deposition collector consisted of a 5-gallon high-density polyethylene bucket with an o-ring sealed lid and a 5-inch-diameter funnel set in the lid. The buckets were filled with 1/2 inch of light mineral oil to prevent evaporation of precipitation between collection intervals. Each funnel contained a tuft of polyfiber filling to prevent clogging of the funnel. The bulk-deposition buckets were collected and replaced with clean buckets every 2 to 4 months. The volume of accumulated water in each sample bucket was measured, and filtered aliquots were collected.

Chloride concentrations of groundwater were estimated from water samples collected at five sites every 2 to 5 months between March 2010 and May 2011 (fig. 3). At the northernmost site, Jinapsan Cave, samples collected from water dripping from the cave ceiling represent water infiltrating through the vadose zone above the lowest groundwater table. Samples from Mataguac Spring represent groundwater that issues from

the perched water body in the volcanic rocks and sediments at Mataguac Hill. Groundwater samples collected from GWA production well Y-15 near Andersen AFB represent groundwater from the major freshwater-lens system in northern Guam. In the south, groundwater samples were collected at Almagosa and Dobo Springs near Fena Valley Reservoir. These springs discharge at the surface contact between a limestone cap and underlying volcanic units, and thus originate from a freshwater body within the older limestone.

All samples were analyzed for chloride using ion chromatography at the USGS National Water Quality Laboratory (NWQL). Chloride concentrations of bulk-deposition samples were determined using NWQL Method I-2058. Chloride concentrations of groundwater samples were determined using NWQL Method I-2057.

Atmospheric Chloride Deposition

Atmospheric chloride deposition at the land surface of Guam varied spatially during the sampling period (table 8). For each measurement period, average daily chloride deposition was calculated as the product of the bulk-deposition chloride concentration and accumulated rainfall volume, divided by the product of the collector catchment area (0.128 ft^2) and measurement period duration. Average daily chloride deposition was greatest at the Jinapsan station, less than a half mile from the coastline on the lowlands at the foot of the northern plateau. Average daily rates of chloride deposition at the three stations on the northern plateau, Y-15, Beng Bing, and Airport, varied by sample location, and were lower than the rates at Jinapsan. Average daily chloride deposition was lowest at the Almagosa station, near Fena Valley Reservoir.

At each bulk-deposition station, the rate of atmospheric chloride deposition also varied during the sampling period. Average daily chloride deposition was highest during measurement period 5 (late-January to late-May), when accumulated rainfall and average daily wind speed were relatively high. In contrast, average daily chloride deposition was relatively low during period 1 (mid-March to mid-June), when rainfall was low, but wind was high. Average daily chloride deposition was also relatively low during period 3 (late August to late October), when rainfall was high, but wind was low. These observed seasonal differences in chloride deposition indicate the importance of measuring chloride deposition for more than just a fraction of a year when using the chloride mass-balance method to estimate recharge on Guam.

Recharge Estimates and Sources of Uncertainty Using Chloride Mass-Balance Method

Recharge was computed for each bulk-deposition station using equation 18 for the Almagosa station and equation 19 for all other stations. C_p at each bulk-deposition site is

Table 8. Summary of chloride concentrations, accumulated rainfall, and chloride deposition at five bulk-deposition stations on Guam, March 2010 through May 2011.

[Station name is the shortened name of the rainfall-collector station used for bulk-deposition sampling; see figure 3 for station locations; Shortened station name (USGS site number): Jinapsan (133825144524101), Y-15 (133318144545702), Beng Bing (133123144513801), Airport (132841144473901), Almagosa (132105144405166); Average daily wind speed is based on wind speed measurements during the measurement period at weather station 914226 at the Guam International Airport; Average chloride deposition is based on each bulk-deposition collector having a catchment area of 0.128 square feet; mi/h, miles per hour; mg/L, milligrams per liter; mg/ft²/d, milligrams per square foot per day; –, no measurements]

Measurement period	Month-day-year		Station name	Average daily wind speed (mi/h)	Chloride concentration (mg/L)	Accumulated rainfall (gallons)	Average chloride deposition (mg/ft²/d)
	From	To					
1	03-11-10	06-13-10	Beng Bing	–	11.7	0.54	1.99
	03-11-10	06-14-10	Airport	11.4	14.8	0.30	1.39
	03-12-10	06-15-10	Almagosa	–	6.83	0.50	1.07
2	06-08-10	08-25-10	Jinapsan	–	11.2	1.59	6.73
	06-13-10	08-24-10	Beng Bing	–	3.42	1.60	2.24
	06-14-10	08-24-10	Airport	8.2	2.48	1.66	1.72
	06-15-10	08-26-10	Almagosa	–	1.76	1.65	1.19
3	08-25-10	10-25-10	Jinapsan	–	5.89	2.01	5.75
	08-25-10	10-22-10	Y-15	–	1.86	1.82	1.72
	08-24-10	10-25-10	Beng Bing	–	2.62	1.17	1.46
	08-24-10	10-22-10	Airport	6.9	1.9	1.57	1.49
	08-26-10	10-23-10	Almagosa	–	1.04	2.08	1.10
4	10-25-10	01-25-11	Jinapsan	–	16.8	1.11	5.99
	10-22-10	01-21-11	Y-15	–	4.63	1.27	1.91
	10-25-10	01-24-11	Beng Bing	–	5.9	0.82	1.57
	10-22-10	01-24-11	Airport	8.6	4.6	0.98	1.42
	10-23-10	01-22-11	Almagosa	–	2.52	1.80	1.47
5	01-25-11	05-26-11	Jinapsan	–	23.7	2.63	15.23
	01-21-11	05-24-11	Y-15	–	6.79	2.10	3.42
	01-24-11	05-25-11	Beng Bing	–	6.86	2.67	4.47
	01-24-11	05-24-11	Airport	10.5	9.58	1.47	3.47
	01-22-11	05-24-11	Almagosa	–	3.89	2.03	1.92

the weighted-average chloride concentration of all samples, where each sample was weighted by the volume of rainfall that accumulated during the measurement period (table 8). This is essentially the same approach used by Orr and others (2002) and Sumioka and Bauer (2003). Mean annual rainfall at each bulk-deposition site was derived from Daly and Halbleib (2006b). C_g is the average chloride concentration of groundwater samples (table 9).

At the Almagosa bulk-deposition station, the weighted-average chloride concentration was 2.60 mg/L. Mean annual surface-water runoff was estimated using runoff records from two nearby continuous stream-gaging stations, 16847000 and 16848500, on the Imong and Maulap Rivers, respectively (fig. 8). The mean annual combined surface-water runoff (*SWR*)

for these two stations is 22.0 in.; mean annual rainfall (*P*), expressed as a uniform depth distributed over the combined drainage areas of these two stations, is 112.1 in. The average chloride concentration of the Dobo and Almagosa Springs samples, 10.2 mg/L, was used for C_g (table 9). Estimated annual recharge is 22.9 in., which is about 20 percent of rainfall, and is lower than the mean annual baseline water-budget model recharge estimate, which is 33.5 in. when uniformly distributed over these drainage basins.

At the Jinapsan bulk-deposition station, the weighted-average chloride concentration was 15.1 mg/L. Average chloride concentration of the water dripping from the ceiling in Jinapsan Cave was 58.1 mg/L. Estimated annual recharge (23.9 in.) is 26 percent of mean annual rainfall (92.1 in.), and

Table 9. Average chloride concentrations of groundwater samples collected between March 2010 and May 2011 at five sites on Guam.

[Site name is the shortened USGS name of the groundwater site; see figure 3 for site locations; mg/L, milligrams per liter]

Site name	USGS site number	Chloride concentration (mg/L)
Jinapsan Cave	133824144524201	58.1
Well Y-15	133318144545701	18.3
Mataguac Spring	133242144530401	14.2
Dobo Spring	132052144405101	10.5
Almagosa Springs	16848000	9.9

Table 10. Range in recharge estimates at three bulk-deposition stations on limestone plateau of northern Guam, computed from the chloride mass-balance method using various groundwater chloride concentrations, and computed using the water-budget model.

[Station name is the shortened USGS station name of the rainfall-collector station used for bulk-deposition sampling; Average chloride concentration is weighted-average of all bulk-deposition samples collected during 2010–2011, where each sample was weighted by the amount of rainfall that accumulated during the measurement period (see table 8); Average annual rainfall is derived from Daly and Halbleib (2006b); Recharge estimate 1 is based on the average chloride concentration of samples from well Y-15 (18.3 mg/L); Recharge estimate 2 is based on the average chloride concentration of samples from Mataguac Spring (14.2 mg/L); Recharge estimate 3 is based on the average chloride concentration used by Ayers (1981) (11.8 mg/L); Water budget is the average annual recharge for subareas in the vicinity of the station, estimated using the water-budget model for baseline conditions; mg/L, milligrams per liter]

USGS site number	Station name	Average chloride concentration (mg/L)	Average annual rainfall (inches)	Recharge estimate (inches per year)			
				1	2	3	Water budget
133318144545702	Y-15	4.53	92.7	22.9	29.6	35.6	46.4
133123144513801	Beng Bing	5.59	94.3	28.8	37.1	44.7	47.1
132841144473901	Airport	5.04	93.1	25.6	33.0	39.8	50.9

is much lower than the mean annual baseline water-budget model recharge estimate for nearby subareas, (51 in).

At the three bulk-deposition stations on the northern plateau, Y-15, Beng Bing, and Airport, recharge estimates are highly dependent on the chloride concentration (C_g) chosen to be representative of groundwater. The weighted-average chloride concentrations for these sites are 4.53, 5.59, and 5.04 mg/L, respectively. Owing to the large differences in the chloride concentrations at the different groundwater sampling sites, a range of recharge estimates at these bulk-deposition stations is computed using the average chloride concentration of samples from well Y-15 (18.3 mg/L) and Mataguac Spring (14.2 mg/L) (table 9). Recharge is also computed at these stations using the average groundwater chloride concentration used by Ayers (1981), which is 11.8 mg/L, and is the average of chloride concentrations from samples collected from three wells in the Yigo-Tumon aquifer sector. The lowest recharge estimates at these stations, those based on well Y-15, range from about 25 to 31 percent of mean annual rainfall (table

10). The middle recharge estimates, those based on Mataguac Spring, range from 32 to 39 percent of rainfall. The highest recharge estimates, those based on the value used by Ayers (1981), range from 38 to 47 percent of mean annual rainfall, and are most similar to the baseline recharge estimate of the water budget model. The middle recharge estimates are most similar to recharge estimated by Ayers (1981), which was 38 percent of annual rainfall. Although each of the rainfall samples listed in Ayers (1981) was collected for only 24 hours, the average chloride concentration of these samples, 4.5 mg/L, is, coincidentally, similar to the weighted-average chloride concentrations determined here.

Whereas groundwater samples represent long-term averages due to mixing and multi-year residence times of water in aquifers, bulk-deposition samples represent atmospheric deposition during the collection period. The interannual variability of atmospheric chloride deposition on Guam is not known, but any variability would influence recharge estimates from the chloride mass-balance method. In particular, the amount of chloride deposited during large storm events such as tropical cyclones is not quantified but is known to be substantial, with past tropical cyclones depositing ocean salt on Guam at concentrations great enough to cause severe browning and defoliation of the vegetation on the island (Guard and others, 1999; Kerr, 2000). Chloride deposition from such large storms could be substantial on Guam because it is in part of the Pacific Ocean frequented by tropical cyclones. Between 1945 and 1997, about 100 tropical cyclones passed within 75 nautical miles of Guam, with half of those being typhoons (Guard and others, 1999). During the sampling period however, no typhoons passed near the island. By not accounting for chloride deposition from large storm events, recharge estimated with the chloride mass-balance method in this study may be low.

Additionally, it was assumed that all chloride in the groundwater samples originated from the atmosphere. Chloride input from septic systems, water-main leakage, or seawater in the aquifer was unlikely at all sampled sites, except at well Y-15 and possibly Jinapsan Cave. Chloride inputs from these sources at these sites would result in higher chloride in groundwater samples and relatively low recharge estimates.

Suggestion for Future Study and Additional Data Collection

The accuracy of recharge calculated using the water-budget model is limited by the quality and availability of data needed to develop model input—land cover, rainfall, irrigation, runoff, soil properties, and evapotranspiration. Initiation of and continued research to characterize these parameters would improve overall confidence in recharge estimates.

Evapotranspiration is an important parameter in any recharge study, yet it is not well-quantified on Guam. Continuing to accurately measure and record ET-related weather parameters at existing weather stations on the island is needed to determine how future climate conditions affect potential ET and recharge compared to historical climate conditions used here. Additional weather stations on Guam would help determine spatial variability of potential ET across the island. Water-budget recharge estimates were most sensitive to two evapotranspiration parameters—canopy evaporation and crop coefficients— that were derived from published studies that were not conducted on Guam. Net-precipitation measurements beneath forested areas, which currently cover about half of the island, would increase confidence in recharge estimates. Perhaps even more important are ET studies that quantify transpiration rates of common trees and vegetation on the island, especially *L. leucocephala* (tangantangan). In the future, new land-cover maps will help assess how land-cover changes affect recharge.

In addition to conducting ET measurements, maintaining a spatially extensive network of rain gages and stream gages is important for estimating the spatial and interannual variability of rainfall and runoff across the island. Incorporating rainfall data from Guam's Doppler weather radar in future water budgets would improve understanding of the spatial variability of rainfall across the island. Monitoring groundwater levels and investigating infiltration rates in the northern half of the island, particularly in closed depressions, may help determine the magnitude of groundwater recharge that can be attributed to fast flow. Maps of the catchment areas and disposal points of storm-drain systems would increase the confidence of recharge estimates in urbanized areas.

For the chloride mass-balance method, the interannual variability of atmospheric chloride deposition is unknown. Quantifying the proportion of chloride deposition to the island that is due to large storms with strong winds and surf, such as those encountered during the nearby passage of tropical cyclones, would help refine the chloride mass-balance recharge estimates reported here.

Summary and Conclusions

In response to population growth, the demand for freshwater on Guam has increased in the past and is likely to continue to increase in the future. Uncertainty in the availability of groundwater resources for increased demand owing to a proposed military relocation to Guam has prompted an investigation of groundwater recharge on the island. This report documents the development of a daily water-budget model for computing groundwater recharge for the entire island of Guam. The model was used to estimate mean annual recharge for various land-cover and rainfall conditions. Recharge was independently estimated using the chloride mass-balance method. Recharge estimates from this study were compared to previous recharge estimates, and the sensitivity of recharge estimates to selected water-budget parameters was evaluated.

Estimated mean annual recharge on Guam is 394.1 million gallons per day (Mgal/d) for baseline conditions, which is 39 percent of mean annual rainfall. Baseline conditions for this study were 2004 land cover and mean annual rainfall during 1961–2005. Compared to rainfall (999.0 Mgal/d) for the entire study area, water inflows from irrigation (0.8 Mgal/d), septic systems (4.8 Mgal/d), and water-main leakage (12.8 Mgal/d) are relatively minor. Of the total water inflow, 13 percent becomes runoff, and 49 percent becomes evapotranspiration (ET) and canopy evaporation.

The total mean annual recharge for the northern aquifer sectors defined by Mink (1991), which encompass most of the northern half of Guam, is 238.0 Mgal/d; this is 51 percent of mean annual rainfall, and is 48.9 inches when expressed as an equivalent depth uniformly spread over the total area of the northern aquifer sectors. Here, of the total water inflow, only 1 percent becomes runoff, and 49 percent becomes ET and canopy evaporation. On a monthly basis, mean recharge is highest in August and is lowest in February.

Recharge is highest in areas that are underlain by limestone, including (1) most of the northern plateau, (2) the older limestone cap in the southern volcanic uplands, (3) the plateau fringing the southeast coastline, (4) Orote Peninsula, and (5) internally drained areas. In these areas where runoff is minor or zero, recharge is typically between 40 and 60 percent of total water inflow. Recharge is relatively high in areas that receive stormwater runoff from storm-drain systems, but is relatively low in urbanized areas where stormwater runoff is routed to the ocean or to other areas. In most of the volcanic uplands in southern Guam, recharge is less than 30 percent of total water inflow.

The water-budget model in this study differs from all previous water-budget investigations on Guam by directly accounting for canopy evaporation in forested areas, quantifying the evapotranspiration rate of each land-cover type, and accounting for evaporation from impervious areas. Recharge estimated in this study is greater than Camp, Dresser & McKee Inc. (1982) and the "minimum" estimates of Mink (1976); similar to Mink (1991), the "probable" estimates of Mink (1976) and two estimates of Habana and others (2009); and less than the estimate of Jocson and others (2002).

For the drought simulation, which used historical rainfall during 1969–73, recharge for the entire island is 259.3 Mgal/d, which is 34 percent lower than recharge computed for baseline conditions. For all northern aquifer sectors defined by Mink (1991), total mean recharge during drought conditions is 162.6 Mgal/d, which is 32 percent lower than mean baseline recharge. For the future land cover water-budget simulation, which represents potential land-cover changes related to the military relocation and population growth, estimated recharge for the entire island is nearly equal to baseline recharge. For the future land-cover scenario with drought conditions, estimated recharge for the entire island is about 33 percent lower than the baseline recharge.

For the entire study area, recharge estimates from the water-budget model are most sensitive to crop coefficients and net precipitation rates. In the southern half of the island, recharge estimates are also sensitive to runoff-to-rainfall ratios.

During March 2010 to May 2011, a series of bulk-deposition samples from five stations on Guam were collected and analyzed for chloride. Additionally, a series of samples from five groundwater sites were collected and analyzed for chloride. Results were used to estimate groundwater recharge using the chloride mass-balance method. Recharge estimates using this method at three bulk-deposition stations on the northern limestone plateau range from about 25 to 47 percent of rainfall. These recharge estimates are similar to the estimate of Ayers (1981) who also used this method. Recharge estimates at each bulk-deposition station, however, are lower than the baseline recharge estimate from the water-budget model used in this study. This may be because no large storms, such as tropical cyclones, passed near Guam during March 2010 to May 2011.

References Cited

Allen, R.G., Pereira, L.S., Raes, Dirk, and Smith, Martin, 1998, Crop evapotranspiration; guidelines for computing crop water requirements: Food and Agriculture Organization of the United Nations, FAO Irrigation and Drainage Paper 56, 300 p.

Asdak, C., Jarvis, P.G., van Gardingen, P., and Fraser, A., 1998, Rainfall interception loss in unlogged and logged forest areas of Central Kalimantan, Indonesia: Journal of Hydrology, v. 206, p. 237–244.

Ayers, J.F., 1981, Estimate of recharge to the freshwater lens of northern Guam: Water and Environmental Research Institute Technical Report no. 21, 20 p.

Beale, C.V., Morison, J.I.L., and Long, S.P., 1999, Water use efficiency of C_4 perennial grasses in a temperate climate: Agriculture and Forest Meteorology, v. 96, p. 103–115.

Bidin, K., and Chappell, N.A., 2004, Sub-canopy rainfall and wet-canopy evaporation in a selectively-logged rainforest, Sabah, Malaysia, in Rahim, N.A., ed., Water Forestry and Land Use Perspectives, International Hydrological Programme, Technical Document in Hydrology no. 70, p. 69–85.

Blaney, H.F., and Criddle, W.D., 1950, Determining water requirements in irrigated areas from climatological and irrigation data: U.S. Department of Agriculture, Division Irrigation and Water Conservation, SCS-TP-96, 48 p.

Bruijnzeel, L.A., and Wiersum, K.F., 1987, Rainfall interception by a young *Acacia Auriculiformis* (A. Cunn) plantation forest in west Java, Indonesia; application of Gash's analytical model: Hydrological Processes, v. 1, p. 309–319.

Calder, I.R., Wright, I.R., and Murdiyarso, D., 1986, A study of evaporation from tropical rain forest—West Java: Journal of Hydrology, v. 89, p. 13–31.

Camp, Dresser & McKee Inc., 1982, Northern Guam Lens Study, Groundwater management program, Aquifer yield report prepared for the Government of Guam: Guam Environmental Protection Agency, variously paged.

Contractor, D.N., and Srivastava, R., 1990, Simulation of saltwater intrusion in the Northern Guam Lens using a microcomputer, Journal of Hydrology, v. 118, p. 87–106.

Contractor, D.N., and Jenson, J.W., 2000, Simulated effect of vadose infiltration on water levels in the Northern Guam Lens Aquifer: Journal of Hydrology, v. 229, p. 232–254.

Crockford, R.H., and Richardson, D.P., 2000, Partitioning of rainfall into throughfall, stemflow and interception; effect of forest type, ground cover and climate: Hydrological Processes, v. 14, p. 2903–2920.

Daly, C., and Halbleib, M., 2006a, Pacific Islands (Guam) average monthly and annual minimum and maximum temperature and mean dewpoint temperature, 1971–2000: PRISM Group at Oregon State University, accessed June 22, 2011, at *http://www.prism.oregonstate.edu/products/pacisl.phtml.*

Daly, C., and Halbleib, M., 2006b, Pacific Islands (Guam) average monthly and annual precipitation, 1971–2000: PRISM Group at Oregon State University, accessed January 21, 2010, at *http://www.prism.oregonstate.edu/products/pacisl.phtml.*

Das, D.K., Singh, Ravinder, and Singh, A.K., 1990, Root water extraction and evapotranspiration by young *Leucaena* and *Eucalyptus* plantations under semi-arid climate: Annals of Agricultural Research, v. 11, no. 1, p. 1–13.

Donnegan, J.A., Butler, S.L., Grabowiecki, Walter, Hiserote, B.A., and Limtiaco, David, 2004, Guam's forest resources, 2002: U.S. Department of Agriculture, Forest Service, Resources Bulletin PNW-RB-243, 32 p.

Dykes, A.P., 1997, Rainfall interception from a lowland tropical rainforest in Brunei: Journal of Hydrology, v. 200, p. 260–279.

Earth Tech, Inc., 1999, Stormwater management/underground injection well closure plan, Andersen AFB—Territory of Guam, prepared for Institute for Environment, Safety, and Occupational Health Risk Analysis Environmental Analysis Division, variously paged.

Engott, J.A., 2011, A water-budget model and assessment of groundwater recharge for the Island of Hawai'i: U.S. Geological Survey Scientific Investigations Report 2011–5078, 53 p.

Engott, J.A., and Vana, T.T., 2007, Effects of agricultural land-use changes and rainfall on ground-water recharge in central and west Maui, Hawai'i, 1926–2004: U.S. Geological Survey Scientific Investigations Report 2007–5103, 56 p.

Fares, Ali, 2008, Water management software to estimate crop irrigation requirements for consumptive use permitting in Hawaii: State of Hawai'i, Commission of Water Resources Management, *http://hawaii.gov/dlnr/cwrm/published reports/PR200808.pdf.*

Fosberg, F.R., 1960, The vegetation of Micronesia, 1. General descriptions, the vegetation of the Mariana Islands, and a detailed consideration of the vegetation of Guam: Bulletin of the American Museum of Natural History, v. 119, no. 1, 76 p., 40 pl.

Gale, M.R., and Grigal, D.F., 1987, Vertical root distributions of northern tree species in relation to successional status: Cana-dian Journal of Forest Research, v. 17, no. 8, p. 829–834.

Giambelluca, T.W., 1983, Water balance of the Pearl Harbor-Honolulu basin, Hawai'i, 1946–1975: University of Hawai'i Water Resources Research Center Technical Report no. 151, 151 p.

Gingerich, S.B., 2003, Hydrologic resources of Guam: U.S. Geological Survey Water-Resources Investigations Report 03–4126, 2 pl.

Guam Environmental Protection Agency, 1997, Individual wastewater disposal systems regulations, at *http://www.guamcourts.org/CompilerofLaws/GAR/22GAR/22GAR002-12.pdf.*

Guam Visitors Bureau, 2010, January 2010 Visitor statistics, at http://visitguam.org/Runtime/GVBResearch.aspx.

Guam Waterworks Authority, 2007a, Guam Waterworks Authority water resources master plan; septic systems and unsewered areas: Guam Waterworks Authority, v. 3, ch. 6, 36 p.

Guam Waterworks Authority, 2007b, Guam Waterworks Authority water resources master plan; water conservation: Guam Waterworks Authority, v. 2, ch. 5, 12 p.

Guam Waterworks Authority, 2007c, Guam Waterworks Authority water resources master plan; population and land use forecast: Guam Waterworks Authority, v. 1, ch. 6, 47 p.

Guard, C.P., Hamnett, M.P., Neumann, C.J., Lander, M.A., and Siegrist, H.G., Jr., 1999, Typhoon vulnerability study for Guam: Water and Environmental Research Institute Technical Report, no. 85, 156 p.

Habana, N.C., Heitz, L.F., Olsen, A.E., and Jenson, J.W., 2009, Vadose flow synthesis for the Northern Guam Lens Aquifer: Water and Environmental Research Institute of the Western Pacific, University of Guam, Technical Report no. 127, 231 p.

Heitz, L., Khosrowpanah, Shahram, and Nelson, Jay, 1997, Sizing of surface water runoff detention ponds: Water and Environmental Research Institute of the Western Pacific, University of Guam, Technical Report no. 80, 74 p.

Hickman, G.C., Vanloocke, Andy, Dohleman, F.G., and Bernacchi, C.J., 2010, A comparison of canopy evapotranspiration for maize and two perennial grasses identified as potential bioenergy crops: Global Change Biology Bioenergy, v. 2, p. 157–168.

Hutjes, R.W.A., Wierda, A., and Veen, A.W.L., 1990, Rainfall interception in the Tai Forest, Ivory Coast; application of two simulation models to a humid tropical system: Journal of Hydrology v. 114, p. 259–275.

ICF International, 2009, Draft North and Central Guam land use plan: Seattle, Wash., ICF International #00824.08, Prepared for Bureau of Statistics and Plans—Government of Guam, 150 p.

Izuka, S.K., Oki, D.S., and Chen, C., 2005, Effects of irrigation and rainfall reduction on ground-water recharge in the Lihue Basin, Kauai, Hawaii: U.S. Geological Survey Scientific Investigations Report 2005–5146, 48 p.

Jackson, I.J., 1975, Relationships between rainfall parameters and interception by tropical forest: Journal of Hydrology v. 24, p. 215–238.

Jackson, R.B., Canadell, J., Ehleringer, J.R., Mooney, H.A., Sala, O.E., and Schulze, E.D.,1996, A global analysis of root distributions for terrestrial biomes: Oecologia, v. 108, no. 3, p. 389–411.

Jocson, J.M.U., Jenson, J.W., Contractor, D.N., 2002, Recharge and aquifer response; Northern Guam Lens Aquifer Guam, Mariana Islands: Journal of Hydrology, v. 260, p. 231–254.

Jones, I.C., and Banner, J.L., 2003, Estimating recharge thresholds in tropical karst island aquifers; Barbados, Puerto Rico and Guam: Journal of Hydrology, v. 278, p. 131–143.

Jordan, C.F., and Heuveldop, J., 1981, The water budget of an Amazonian rain forest: Acta Amazonica, v. 11, p. 87–92.

Kerr, A.M., 2000, Defoliation of an island (Guam, Mariana Archipelago, Western Pacific Ocean) following a saltspray-laden 'dry' typhoon: Journal of Tropical Ecology, v. 16, p. 895–905.

Kubota, Hisayuki, and Wang, Bin, 2009, How much do tropical cyclones affect seasonal and interannual rainfall variability over the Western North Pacific?: Journal of Climate, v. 22, p. 5495–5510.

Lander, M.A., 1994, Meteorological factors associated with drought on Guam: Water and Environmental Research Institute of the Western Pacific, University of Guam, Technical Report no. 75, 39 p.

Lander M.A., and Guard, C.P., 2003, Creation of a 50-year annual rainfall database, annual rainfall climatology, and annual rainfall distribution map for Guam: Water and Environmental Research Institute of the Western Pacific, University of Guam, Technical Report no. 102, 20 p.

Lander, M.A., Jenson, J.W., and Beausoliel, Colette, 2001, Responses of well water levels on northern Guam to variations of rainfall and sea level: Water and Environmental Research Institute of the Western Pacific, University of Guam, Technical Report no. 94, 36 p.

Lloyd C.R., Gash, J.H.C., Shuttleworth, W.J., and Marques F, A. de O., 1988, The measurement and modelling of rainfall interception by Amazonia rain forest: Agricultural and Forest Meteorology, v. 43, p. 277–294.

Maurer, D.K., Berger, D.L., and Prudic, D.E., 1996, Subsurface flow to Eagle Valley from Vicee, Ash, and Kings Canyons, Carson City, Nevada, estimated from Darcy's law and the chloride-balance method: U.S. Geological Survey Water-Resources Investigations Report 96–4088, 74 p.

McJannet, David, Wallace, Jim, Fitch, Peter, Disher, Mark, and Reddell, Paul, 2007, Water balance of tropical rainforest canopies in north Queensland, Australia: Hydrological Processes, v. 21, p. 3473–3484.

Meijer, Arend, Reagan, Mark, Ellis, Howard, Shafiqullah, Muhammad, Sutter, John, Damon, Paul, and Kling, Stanley, 1983, Chronology of volcanic events in the eastern Philippine Sea, in Hayes, D.E., ed., The tectonic and geologic evolution of southeast Asian seas and islands; part 2: Washington, D.C., American Geophysical Union, Geophysical Monograph 27, p. 349–359.

Mink, J.F., 1976, Groundwater resources of Guam; Occurrence and development: Water and Environmental Research Institute of the Western Pacific, University of Guam, Technical Report no. 1, 276 p.

Mink, J.F., 1991, Groundwater in northern Guam; Sustainable yield and groundwater development: Barrett Consulting Group for the Public Utility Agency of Guam.

Moran, D.C., and Jenson, J.W., 2004, Dye trace of groundwater flow from Guam International Airport and Harmon Sink to Agana Bay and Tumon Bay, Guam: Water and Environmental Research Institute of the Western Pacific, University of Guam, Technical Report no. 97, 39 p.

National Oceanic and Atmospheric Administration, 2009, C-CAP land cover, Territory of Guam 2005, accessed March 16, 2010, at *http://www.csc.noaa.gov/digitalcoast/data/ccaphighres/download.html.*

National Renewable Energy Laboratory, 2007, Digital solar radiation data for the United States and its territories, accessed February 25, 2010, at *http://rredc.nrel.gov/solar/old_data/nsrdb/.*

Ogden Environmental and Energy Services, Inc., 1995, Site investigation report for drainage basin holding pond, Naval Air Station Agana Guam, Mariana Islands: Comprehensive Long-Term Environmental Action Navy I (CLEAN I) Program, Contract Task Order (CTO) No. 0036, April 1995, variously paged.

Oki, D.S., 2002, Reassessment of ground-water recharge and simulated ground-water availability for the Hāwī area of North Kohala, Hawaii: U.S. Geological Survey Water-Resources Investigations Report 02–4006, 62 p.

Oki, D.S., 2008, The significance of accounting order for evapotranspiration and recharge in monthly and daily threshold-type water budgets: U.S. Geological Survey Scientific Investigations Report 2008–5163, 11 p.

Orr, L.A., Bauer, H.H., and Wayenberg, J.A., 2002, Estimates of ground-water recharge from precipitation to glacial-deposit and bedrock aquifers on Lopez, San Juan, Orcas, and Shaw Islands, San Juan County, Washington: U.S. Geological Survey Water-Resources Investigations Report 02–4114, 113 p.

Prych, E.A., 1995, Using chloride and chlorine-36 as soil-water tracers to estimate deep percolation at selected locations on the U.S. Department of Energy Hanford site, Washington: U.S. Geological Survey Open-File Report 94–514, 125 p.

Reagan, M.K., and Meijer, Arend, 1984, Geology and geochemistry of early arc-volcanic rocks from Guam: Geological Society of America Bulletin, v. 95, p. 701–713.

Roupsard, O., Bonnefond, Jean-Marc, Irvine, M., Berbigier, P., Nouvellon, Y., Dauzat, J., Taga, S., Hamel, O., Jourdan, C., Saint-André, L. Miale-Serra, I., Labouisse, Jean-Pierre, Epron, D., Joffre, R., Braconnier, S., Rouzière, A., Navarro, M., and Bouillet, Jean-Pierre, 2006, Partitioning energy and evapo-transpiration above and below a tropical palm canopy: Agricultural and Forest Meteorology, v. 139, p. 252–268.

Savenije, H.H.G., 2004, The importance of interception and why we should delete the term evapotranspiration from our vocabulary: Hydrological Processes, v. 18, p. 1507–1511.

Scanlon, B.R., 1991, Evaluation of moisture flux and chloride data in desert soils: Journal of Hydrology, v. 128, p. 137–156.

Schellekens, J., Scatena, F.N., Bruijnzeel, L.A., Wickel, A.J., 1999. Modelling rainfall interception by a lowland tropical rain forest in northeastern Puerto Rico: Journal of Hydrology 225, p. 168–184.

Scholl, M.A., Ingebritsen, S.E., 1995, Total and non-seasalt sulfate and chloride measured in bulk precipitation samples from the Kilauea volcano area, Hawaii: U.S. Geological Survey Water-Resources Investigation Report 95–4001, 32 p.

Scholl, M.A., Ingebritsen, S.E., Janik, C.J., and Kauahikaua, J.P., 1995, An isotope hydrology study of the Kilauea volcano area, Hawaii: U.S. Geological Survey Water-Resources Investigation Report 95–4213, 44 p.

Scholl, M.A., Ingebritsen, S.E., Janik, C.J., and Kauahikaua, J.P., 1996, Use of precipitation and groundwater isotopes to interpret regional hydrology on a tropical volcanic island; Kilauea volcano area, Hawaii: Water Resources Research, v. 32, no. 12, p. 3525–3537.

Shuttleworth, W.J., 1993, Evaporation, chap. 4 of Maidment, D.R., ed., Handbook of hydrology: New York, McGraw-Hill, p. 4.1–4.53.

Siegrist, H.G., Jr., Reagan, M.K., Randall, R.H., and Jenson, J.W., 2008, Geologic map and sections of Guam, Mariana Islands: Water and Environmental Research Institute of the Western Pacific, University of Guam, map, available at http://www.weriguam.org/pdf/geologic-maps/map-of-guam.pdf.

Sumioka, S.S., and Bauer, H.H., 2003, Estimating ground-water recharge from precipitation on Whidbey and Camano Islands, Island County, Washington, water years, 1998 and 1999: U.S. Geological Survey Water-Resources Investigations Report 03–4101, 49 p.

Suyker, A.E., and Verma, S.B., 2009, Evapotranspiration of irrigated and rainfed maize-soybean cropping systems: Agriculture and Forest Meteorology, v. 149, p. 443–452.

Taborosi, Danko, 2006, Karst inventory of Guam, Mariana Islands: Water and Environmental Research Institute of the Western Pacific, University of Guam, Technical Report no. 112, 141 p., ill.

Taylor, C.J., and Nelson, H.L., Jr., , 2008, A compilation of provisional karst geospatial data for the Interior Low Plateaus physiographic region, central United States: U.S. Geological Survey Data Series 339, appendix 1.

Thornthwaite, C.W., and Mather, J.R., 1955, The water balance: Publications in Climatology, v. 8, no. 1, p. 1–104.

Tracey, J.I., Jr., Schlanger, S.O., Stark, J.T., Doan, D.B., and May, H.G., 1964, General geology of Guam: U.S. Geological Survey Professional Paper 403–A, 104 p., 3 pl.

U.S. Army Corps of Engineers, 1980, Guam storm drainage manual; A technical report from the comprehensive study of Guam's water and related land resources, variously paged.

U.S. Census Bureau, 2011, Table 1313; Estimated resident population with projections: U.S. Census Bureau, Statistical Abstract of the United States; 2010, Puerto Rico and the Island Areas, accessed May 19, 2011, at http://www.census.gov/compendia/statab/cats/puerto_rico_the_island_areas.html.

U.S. Department of Agriculture, 2006a, Detailed vegetation map for the Island of Guam: U.S. Department of Agriculture, Forest Service Region 5, State and Private Forestry, Forest Health Protection, accessed March 29, 2010, at http://www.fs.fed.us/r5/spf/fhp/fhm/landcover/islands/index.shtml.

U.S. Department of Agriculture, 2006b, Orthophoto high resolution mosaic for Guam: U.S. Department of Agriculture, Natural Resources Conservation Service.

U.S. Department of Agriculture, 2009, Soil survey geographic (SSURGO) database for Territory of Guam: U.S. Department of Agriculture, Natural Resources Conservation Service, accessed March 31, 2010 at *http://soildatamart.nrcs.usda.gov/Survey.aspx?County=GU010.*

U.S. Navy, Naval Facilities Engineering Command, Pacific [NAVFAC Pacific], 2010a, Executive summary, *in* Final environmental impact statement, Guam and CNMI military relocation: Pearl Harbor, Hawaii, NAVFAC Joint Guam Program Office, p. ES–1 to ES–60, available at *http://www.guambuildupeis.us/final_documents.*

U.S. Navy, Naval Facilities Engineering Command, Pacific [NAVFAC Pacific], 2010b, Final storm water implementation plan for the Guam road network, *in* Final environmental impact statement, Guam and CNMI military relocation: Pearl Harbor, Hawaii, NAVFAC Joint Guam Program Office, v. 9, appendix G, ch. 5, p. 209–334, available at *http://www.guambuildupeis.us/final_documents.*

U.S. Navy, Naval Facilities Engineering Command, Pacific [NAVFAC Pacific], 2010c, Marine Corps relocation—Guam, Proposed action and alternatives, *in* Final environmental impact statement, Guam and CNMI military relocation: Pearl Harbor, Hawaii, NAVFAC Joint Guam Program Office, v. 2, ch. 2, p. 2–1—2–160, available at *http://www.guambuildupeis.us/final_documents.*

U.S. Navy, Naval Facilities Engineering Command, Pacific [NAVFAC Pacific], 2010d, Related actions—Utilities and roadway projects, Proposed action and alternatives, Potable water, *in* Final environmental impact statement, Guam and CNMI military relocation: Pearl Harbor, Hawaii, NAVFAC Joint Guam Program Office, v. 6, ch. 2.2, p. 2–26—2–50, available at *http://www.guambuildupeis.us/final_documents.*

Viessman, Warren, Jr., and Lewis, G.L., 2003, Introduction to hydrology (5th ed.): Upper Saddle River, N.J., Prentice Hall, 612 p.

Wahl, K.L., and Wahl, T.L., 1995, Determining the flow of Comal Springs at New Braunfels, Texas: Proceedings of Texas Water '95, A component conference of the American Society of Civil Engineers International Conference on Water Resources Engineering, 1st, San Antonio, Texas, 1995, p. 77–86.

Wallace, Jim, and McJannet, Dave, 2006, On interception modelling of a lowland coastal rainforest in northern Queensland, Australia: Journal of Hydrology, v. 329, p. 477–488.

Ward, P.E, Hoffard, S.H., and Davis, D.A., 1965, Hydrology of Guam, *in* Geology and hydrology of Guam, Mariana Islands, U.S. Geological Survey Professional Paper 403–H, 28 p., 1 pl.

Young, F.J., 1988, Soil survey of Territory of Guam: U.S. Department of Agriculture, Soil Conservation Service, 166 p., maps.

This page intentionally left blank.

Produced in the Western Region, Menlo Park, California
Manuscript approved for publication, January 27, 2012
Edited by Iris Collies and James W. Hendley II
Layout by Judy Weathers

Johnson—A Water-Budget Model and Estimates of Groundwater Recharge for Guam—Scientific Investigations Report 2012–5028

≥USGS

www.ingramcontent.com/pod-product-compliance
Lightning Source LLC
Chambersburg PA
CBHW081853170526
45167CB00007B/3005